CURATED DECAY

Curated Decay

HERITAGE BEYOND SAVING

Caitlin DeSilvey

University of Minnesota Press
Minneapolis • London

An earlier version of chapter 2 was published as "Observed Decay: Telling Stories with Mutable Things," *Journal of Material Culture* 11 (2006): 318–38; http://dx.doi.org/10.1177/1359183506068808; reprinted with permission of Sage Publications. Portions of chapter 4 were published as "Palliative Curation: Art and Entropy on Orford Ness," in *Ruin Memories: Materialities, Aesthetics, and the Archaeology of the Recent Past*, ed. Bjørnar Olsen and Þóra Pétursdóttir, 79–91 (London: Routledge, 2014); reprinted by permission of Taylor & Francis Group.

All illustrations are photographs by the author.

Published by the University of Minnesota Press
111 Third Avenue South, Suite 290
Minneapolis, MN 55401-2520
http://www.upress.umn.edu

The University of Minnesota is an equal-opportunity educator and employer.

Library of Congress Cataloging-in-Publication Data
DeSilvey, Caitlin, author.
Curated decay : heritage beyond saving / Caitlin DeSilvey.
Minneapolis : University of Minnesota Press, 2017. | Includes bibliographical references and index.
Identifiers: LCCN 2016015344| ISBN 978-0-8166-9436-5 (hc) | ISBN 978-0-8166-9438-9 (pb)
Subjects: LCSH: Historic preservation—Philosophy. | Cultural property—Protection. | Historic sites—Conservation and restoration.
Classification: LCC CC175.D465 2017 | DDC 363.6/9—dc23
LC record available at https://lccn.loc.gov/2016015344

Contents

Postpreservation

LOOKING PAST LOSS

> All the new thinking is about loss.
> In this it resembles all the old thinking.
>
> *Robert Hass, "Meditation at Lagunitas"*

A SOLITARY CHIMNEY STACK rises from a scrubby patch of open ground at the northern edge of our Cornish village. The tapered column of granite and brick pierces the horizon, as it has since the middle of the nineteenth century, when it was built. It has become so seamlessly stitched into the backdrop of village life that no one really notices it any more. When it was constructed, the chimney was attached to a masonry building housing a massive steam engine that pumped water out of mine shafts driven deep into the hill. The ground around the structure was a busy industrial yard; the slopes below, now woodland, were loud with the grinding of waterwheels and ore crushers, and with the trundling of carts that hauled granite from the quarry at the head of the valley to a quay below on the Helford River. These sounds are silenced now, and the scrubland around the remnant chimney is choked with brambles, nettles, gorse, and buddleia— but also sloe, wild roses, and poppies.

Although the chimney appears solid enough from a distance, on inspection its advanced age becomes apparent. The mortar in the granite rubble stonework is friable and loose;

overhanging masonry sections jut out along the seam where the engine house wall once attached, and only thick nets of ivy appear to hold them in place. The shaft inside is packed with branches and sticks deposited by generations of jackdaws. The integrity of the structure appears to be symbiotic, a weave of root and rock, rather than singular. One day when I was poking about in the ruins of the structure's crumbled flank I found a small stoneware bottle. It stood about three and a half inches high, with straight sides and a flared rim, dingy and clogged with soil. At home, I cleaned it to reveal an uneven orange glaze and the mark of a potter's thumb on its shoulder.

The chimney itself can also be understood as a vessel of sorts, holding material memories of the industrial past in this place. Depending on how you look at it, however, in its current state the chimney is either half empty or half full. The school of half empty would see the chimney as a threatened object, its significance gradually eroding as its condition deteriorates. English Heritage (the U.K. public body responsible for the national system of heritage protection) listed the chimney as a feature of "special architectural or historic interest" in 1988, but it is privately owned, and no active measures have been taken to stabilize it.[1] If, at some point in the future, someone happened to notice that the chimney is near collapse, it is likely that proposals would come forward to save and secure it—to strip off the ivy, repoint the mortar, clear the base, and install an informational plaque with a potted history to justify the expense of intervention. The structure would be infilled with official memory and asked to perform as an object of heritage.

Another way of looking at the chimney in its current limbo state would posit that the feature's ongoing decay, rather than threatening to hollow out the memory and the meaning of the structure, instead has its own productive relation to the past. Since the chimney stopped performing its original function at the end of the nineteenth century, it has been caught up in a variety of processes, from opportunistic salvage of the engine house stone for use in other building projects to gradual, spontaneous colonization by equally opportunistic plant and animal species. It is possible to see a fullness in the current state of the structure as it sheds one arrangement of matter to adopt another. As the chimney becomes less legible as an object of industrial heritage, it becomes possible to read other narratives out of its remains, to trace the granite blocks from their source in the Cornubian batholith to their temporary enrollment in this structure, to follow the ivy roots into the seams of the stone to learn how they find nourishment in mineral mortar, to envision a future in which the chimney no longer stands but something of its substance and its story persists nonetheless—to understand change not as loss but as a release into other states, unpredictable and open.

We live in a world dense with things left behind by those who came before us, but we only single out some of these things for our attention and care. We ask certain buildings, objects, and landscapes to function as mnemonic devices, to remember the pasts that produced them, and to make these pasts available for our contemplation and concern. The language that we use when an object or structure is recognized for its potential contribution to cultural memory work immediately presumes a threat, a risk of loss.[2] We speak of vulnerable places and things needing protection, conservation, and preservation. Action is required to restore or maintain the physical integrity of the threatened object and ensure its survival. Intervention and treatment aim to protect things from outright destruction or neglect as well as more indirect processes of erosion, weathering, decay, and decomposition. But what happens if we choose not to intervene? Can we uncouple the work of memory from the burden of material stasis? What possibilities emerge when change is embraced rather than resisted?

Although in present-day Euro-American heritage contexts such questions have a whiff of heresy, it has not always been so. The prevailing preservation paradigm, which declares that certain objects must be retained for the benefit of future generations and asserts the moral imperative of material conservation, only emerged in the late nineteenth century.[3] As part of a broader cultural shift toward the disciplining of knowledge and expertise, objects of presumed historic value became subject to new standards of classification, recording, and documentation.[4] Once safely contained within schedules, lists, and inventories, artifacts and structures fell under the presumption of protection. Graham Fairclough writes, "The obsession with physical conservation became so embedded in twentieth century mentalities that it is no longer easy to separate an attempt to understand the past and its meaning from agonising about which bits of it to protect and keep. . . . The remains of the past . . . seem to exist only to be preserved."[5] A rash of legislation in the early part of the twentieth century secured expectations that all reasonable attempts would be made to protect designated entities in perpetuity.[6] Other perspectives, more accommodating and appreciative of material transience and change, were silenced or sidelined.

In recent years, some have called for a reevaluation of our commitment to perpetual material protection. "Our heritage system is constipated," argues Maria Balshaw. "It is time for a no-blame conversation about letting some things change and even letting some things go."[7] Rodney Harrison writes of a "crisis of accumulation" in heritage practice and the need to make "active decisions to delist or cease to conserve particular forms of heritage," lest we become overwhelmed.[8] Some scholars frame the "forgetting" catalyzed by acts of deliberate deaccession as an essential constituent of a dynamic and productive relation to the present and the future. Mark Augé observes, "We must forget in order to remain present, forget in order not to die, forget in order to remain faithful."[9] Others, however, have been keen to point out that cultural amnesia does not necessarily follow from material erasure, and encroaching absence may paradoxically facilitate the persistence of memory and significance.[10] Þóra Pétursdóttir, in her work on disused Icelandic herring fishing stations, observes that abandonment can be understood as termination, or "an evolving and dynamic context in its own right."[11] In his discussion of the destruction of a twelfth-century Norwegian church, Cornelius Holtorf asserts that processes of change and creative transformation may actually help maintain a connection to the past rather than sever it.[12] It is possible to perform remembrance through transience, although this may require a willingness to find value in alternative material forms.

In this book, I explore the implications of a set of unorthodox premises: the disintegration of structural integrity does not necessarily lead to the evacuation of meaning; processes of decay and disintegration can be culturally (as well as ecologically) productive; and, in certain contexts, it is possible to look beyond loss to conceive other ways of understanding and acknowledging material change.

Each chapter in this book considers a site where exploration of alternatives to material conservation has been deliberate and considered (rather than a post hoc rationalization of benign neglect). I should make it clear that I am primarily interested in places where original function has given way to post-production recognition of historic value. I visit the Montana homestead where these ideas first took root; a nineteenth-century Cornish harbor; a remote Cold War research complex; a postindustrial landscape park; a modernist Scottish seminary; a derelict gunpowder works; an abandoned mining camp; and an imperilled lighthouse.[13] In each of these places, I try to explain the thinking that informed the decision (or intention) to reserve repair and defer maintenance. In some of these places, decay has been allowed to run its course out of an appreciation for its aesthetic effects. Other sites were ceded to allow natural process to return to a previously managed landscape. Underlying these philosophical grounds are, inevitably, pragmatic considerations about the availability of resources and the feasibility of continued investment. What characterizes each site, however, is some form of improvisation and innovation in the face of uncertainty.[14] In each place, I trace the tangle of why and how, and I try to extend the potential for doing things otherwise, for acknowledging (historic) significance without arresting process, by proposing my own experiments with curatorial and interpretive practice. Each chapter presents a snapshot in time, a discrete interval that is, by necessity, truncated midstory. All of these places are caught up in currents of continued change, and this means that by the time this book appears in print, they will have moved on—either to be drawn back into the "safe" harbor of heritage protection or to pass further over the threshold into accelerated decline.

In order to describe what is happening in these perforated places, I need to draw on new ways of storying matter—surfacing meaning that extends beyond cultural frames of reference, and inviting in other agencies and other narrative forms. I locate my analysis in the fine grain of materials, where interpretation stitches down the ragged line between presence and absence, here and gone, object and process. In the telling, an inevitable tension arises between my desire to understand and articulate the intentions of the human subjects responsible for these places and my simultaneous interest in identifying expressions of material and ecological sovereignty. I describe the metamorphosis of the material fabric in these places and tease out the stories that are generated through processes of colonization, dissolution, and disintegration. To borrow a phrase from Jane Bennett, I come to these places with an "anticipatory readiness . . . a perceptual style congenial to the possibility of thing power."[15] In my desire to be as precise as possible about the processes I observe at work, I am often forced to draw on bodies of knowledge that are outside my expertise—ecology, chemistry, materials science. I may risk failure or misinterpretation, but I seek reassurance in the awareness that potent moments always involve some form of perplexity, a recognition that forces beyond my ken are at work and that all I can do is describe what I see within the limits of my understanding.

I take heart from other thinkers who accept that there are worlds that lie beyond the borders of our ability to articulate them. Bennett writes of our encounter with a world of "entities not entirely reducible to the contexts in which (human) subjects set them, never entirely exhausted by their semiotics."[16] Luke Introna proposes that we allow ourselves to be affected by forms and substances that we do not attempt to control or order, cultivating "an affective mode of comportment towards the other that refuses to turn the becoming of the other into containable things or wholes."[17] Associated with this cultivation of openness and uncertainty is a reluctance to rely on notions of nature or culture as stable categories to which objects can be intuitively allocated.[18]

negative terms. Two distinct but related terms—entropy and decay—have particular relevance for the discussion I unfold in the following chapters, and it is worth spending some time unpacking them here.

Rudolf Clausius coined the term "entropy" from the Greek *entropein*, "transformation and change."[28] Outside of the disciplines in which it functions as a working concept (information theory, statistical mechanics, physics), reference to entropy is usually a shorthand invocation of a state of increasing disorder, chaos, or disorganization. Although definitions of entropy vary widely depending on the context in which they are applied, most scholars who use the term in their work would agree that the emphasis on disorder is misleading; entropy is more accurately defined as a measure of the multiplicity of potential arrangements of matter within a given system. Systems with a greater range of potential configurations are described as existing in a state of high entropy. For example,

> A tidy or ordered room is a room where the items in the room inhabit a small set of possible places—the books on the bookshelf, the clothes in the dresser, and so on— while a messy or disordered room is the set of all other configurations. . . . Thus, a messy room does not have higher entropy because it is messy or disordered but rather because there are more configurations [that would count as messy] than an ordered or tidy room. That is, its multiplicity is higher.[29]

In heritage contexts, a consolidated or conserved structure expresses a limited set of potential configurations (paint on trim, masonry pointed, roof in true); a structure that is caught up in active processes of decay and dereliction has many more. The multiplicity that results when maintenance and repair is withheld is the measure of entropy in the structure, and it is inherently unpredictable and uncertain. Another standard definition of entropy holds it to be the amount of energy in a physical system that cannot be used to do work. In the systems this book is concerned with, "work" is allied to the work of memory. Massive amounts of energy are invested to keep heritage systems in a steady state so that the matter contained within them will continue to function as a cultural mnemonic device. Such work can involve freezing, irradiation, treating for mold, inserting borate rods, and any number of other preventive and protective techniques. In an entropic system, however, matter continually degrades, energy is lost, and an element of chance enters into the equation.

Perhaps, as some have suggested, entropy can best be described as possibility, rather than through reference to chaos and disorder: "Entropy is an additive measure of the number of possibilities available to a system. . . . As the constraints that inform a living organism dissolve, the entropy of the organism increases. . . . Yet even in the death, new possibilities are sown."[30] In its biological expression, as noted above, entropy is closely aligned with decay. Decay occurs when a complex of biological, chemical, and physical processes—each driven by specific agents and elements—combines to break down the integrity of a substance and to make its components available for enrollment in other projects. The decomposition catalyzed by enzymes and microorganisms, for example, releases nutrients and increases the fertility of surrounding substrates, allowing for the emergence of new forms of growth. As Jane M. Jacobs and Stephen Cairns point out, "Biological and ecological concepts of decay are full of activity, exchange, acquisition and redistribution. Decay is as life-giving as it is life-taking."[31] There is a whole field devoted to the study of the biodeterioration of cultural heritage, but the focus of scholarship, for the most part, remains resolutely fixed on the destructive aspects of the decay process and on identifying strategies for protection and remediation.

Both decay and entropy carry potentially contradictory meanings, and depending on context they are inflected as negative or positive, generative or destructive. In relation to built structures and artifacts, decay is usually framed either through a "logic of loss" or a logic of renewal and rebirth.[32] Land artist Robert Smithson identified the "clashing aspect of the entropic tendency," which he defined as an irreconcilable tension between different perceptions and valuations of entropic process.[33] This tension was embedded within his own thinking, in that he sometimes described the "entropic mood" as a gradual collapse of culture toward the banal, the empty, and the vapid.[34] More often, Smithson asserted that a willingness to "recognise the entropic condition rather than try to reverse it" could generate positive reformulations and catalyze the continual remaking of matter and culture.[35] In one essay, Smithson cites physicist P. W. Bridgman: "Like energy, entropy is in the first instance a measure of something that happens when one state is transformed into another."[36] Jeremy Till has written about Smithson's collaboration with entropic process as a signature feature of works such as *Spiral Jetty*, a spiral of rocks reaching into Utah's Great Salt Lake, which is "at the same time natural/artificial, of the land/of the water, stable/decaying."[37]

What does any of this have to do with the prosaic practice of heritage management? As Gavin Lucas reminds us, "entropy is a social as well as a natural phenomenon," and our handling of the material record that has persisted from the past into the present is always a negotiation of the "virtual extremes of total preservation and total erasure."[38] A focus on entropy allows us to look to the processes by which worlds are assembled and to accept that any given system, be it a granite chimney stack or an artwork, has the potential to unfold along multiple trajectories; what may appear as erasure on one register may be generative of new information on another. An attentive relation to material systems and their histories involves following trajectories of change and transformation rather than arresting them.

Of course, such an experimental heritage practice is at odds with conventional framings of the relationship between the material past and the memorial present. Objects of heritage are preserved, most transparently, in order to stabilize memory in material form and to stabilize associated identity formations.[39] At the scale of the collective, acts of preservation and designation enroll certain structures and artifacts to function as mnemonic anchors.[40] The memories associated with these monumental forms may be popular or elite, consensual or contested, but the link between material persistence and memorial function goes largely unquestioned.[41] On an intimate register, people use objects as memory prompts to materialize elements of identity and experience.[42] Conservation of the material past, in its most familiar mode, is an act of "self-preservation," an impulse that seeks to maintain the relation between self and surround.[43] While it is possible to make an intellectual or aesthetic argument for postpreservation heritage practice, such a proposal presents a fundamental challenge at the base level of self. The act of "saving" implicates us, as individuals, in the biography of an artifact—or, as some have suggested, we save things not "because they are valued, but rather they are valued because they are being saved."[44] With each act of preservation, the vulnerable object becomes (a little bit of) us, and its unmaking threatens to unmake our identities as well.

The transitive model described above links materiality, memory, and subjectivity through mutually reinforcing
chains of reference. This model relates awkwardly to materials that are caught up in processes of change and transformation. Such materials yield their significance more readily
when memory is framed as generative rather than transitive, a
"culturally mediated material practice that is activated by embodied acts and semantically dense objects."[45] Objects with
mnemonic resonance contribute their own resources and po-
tentialities to an encounter, and these may exceed our ability
to contain or comprehend them.[46] If memory is understood
not as something that is deposited within material containers
for safekeeping but as something that is "ignited in dialogue
between mind and matter," then it does not necessarily need
to rely on a stable material form for its expression.[47]

In the interface between materiality and sociality, different agencies—discursive and practiced, textual and tactile—
may contribute to the production of memory. Remembrance
in this mode involves a willingness to accept the unsettling
of our sense of ourselves as autonomous agents and to think
instead about the work of assembling meaning as a collaboration with an array of other materials, forces, and organisms.
In this more dispersed and fluid understanding of subjectivity, materiality is not a static field of reference that awaits in-
scription from an active mind but is itself constitutive of (new
forms of) human selfhood, as distributed through intimate relations with other entities—plants, stones, dust.[48] With regard
to heritage objects, such a shift in thinking requires a more
nuanced appreciation of the forces that lead to forgetting—
acts of preservation obscure and eliminate certain traces of
the past even as they secure others. It may be that in some
circumstances a state of gradual decay provides more oppor-

tunities for memory making, and more potential points of engagement and interpretation, than the alternative.[49]

The potential to uncouple memory work from material stability—to question, as Aron Vinegar and Jorge Otero-Pailos have suggested, the "primacy granted to presence and materiality in preservation"—is the subject of this book.[50] The stories I tell, however, end up being as much about holding together as they are about coming apart. In each of the places I visit, I acknowledge the anxiety associated with surrender, with allowing processes of change to progress unchecked. It goes against the grain of human nature to step back and allow things to collapse; the urge to step in at the last minute to avert material disintegration is a powerful one.[51] For this reason, much of my discussion ends up being about the inevitability of intervention and the limits to radical innovation. Some of these limits are subjective, but many more are structural; "protection" in heritage contexts applies not just to the physical form of discrete objects and structures but to those who own them and those who encounter them. A thicket of laws and policies are intended to protect owners' liability and to protect publics from exposure to dangerous substances and risky situations. Sanctioned inaction is difficult to accommodate within existing regulatory frameworks, and in certain contexts the approaches I describe here would be entirely inappropriate—as well as illegal. In many of the sites that I discuss in this book, the laws that set expectations for the protection of built heritage also come into uneasy contact with legislation that applies to the management of ecosystems, as the opportunistic organisms that are the first to take root in abandoned sites and structures are frequently subject to control as invasive species. In other places, ecological arguments bolster management positions with regard to cultural heritage objects, and invocations of "natural process" and "managed decline" play into other agendas and interests. There is a politics as well as a poetics to the approaches I introduce here, and much of the detail in the following chapters lies in my articulation of extended negotiations over who ultimately has the power to decide when to do things otherwise, and why.

The insights that I share in this book are aligned with a wider cultural recognition that we need to find ways to inhabit change rather than deny or deflect it, and to find meaning in transition, transience, and uncertainty.[52] If one accepts that we live in a world of ecological unraveling and rising seas, fragile economies and gathering storm clouds, then one is forced to admit that we may not be in control anymore, if we ever were. When Ernest Callenbach, the author of the 1975 novel *Ecotopia*, died in 2012, he left behind a document on his computer that included this prescient observation:

> Humans tend to try to manage things: land, structures, even rivers. We spend enormous amounts of time, energy and treasure in imposing our will on nature, on pre-existing or inherited structures, dreaming of permanent solutions, monuments to our ambitions and dreams. But in periods of slack, decline or collapse, our abilities no longer suffice for all this management. We have to let things go. All things *go* somewhere: they evolve, with or without us, into new forms. So as the decades pass we should try not always to futilely fight these transformations. . . . We can embrace this process of devolution: embellish it when strength avails, learn to love it.[53]

As Callenbach's comment suggests, when protection can no longer be sustained at the levels we have become accustomed to, we will need new ways of making sense of the world and our relationship to it. While in one sense this observation may seem fatalistic, willing to prematurely accept an impending upheaval (that may or may not materialize), I would argue that the transformation of our relation to the material past is both a necessity and an opportunity. As I hope will become clear in this book, I am not advocating a position of acquiescence and indifference in the face of change. I am trying to muster the cultural and practical resources that will be required to think about process and transformation as openings, invitations to engagement and experimentation. We need ways of valuing the material past that do not necessarily involve accumulation and preservation—ways that instead countenance the release of some of the things we care about into other systems of significance.

A couple of years ago I had the opportunity to speak to an audience in Glasgow about my research. After listening to my presentation about the intentional accommodation of ruination at a Cold War military site, a writer friend commented, "This is either an incredibly old theme or an incredibly new one." Yes. A fascination with things ruined, decayed, derelict, and transient plays out on a continuous loop in Western aesthetic and intellectual traditions, inflected through each iteration with a slightly different emphasis, each meeting a different need. This inheritance forms a kind of undertow to the work I want to do in this book, an insistent tug that asserts the continued relevance of these older ways of seeing, now bundled into post hoc structures of feeling—Baroque, Gothic, Romantic, Picturesque. The labels signify cultural moments when people saw something of value in material transformation and disorder, rather than stasis; when sensibility was attentive to transience and titillated by decay.

Although these traditions are clearly relevant to the work I set out to do here, and I return to some of these precedents in the chapters that follow, my aim in this book is to try to articulate a way of relating to disarticulating places and things that exposes new possibilities for engagement and interpretation rather than reinventing inherited ones.[54] It may be that, as Robert Hass points out, "All the new thinking is about loss. / In this it resembles all the old thinking," and that it is impossible to disassociate my argument from its weighty antecedents.[55] But I want to try, and while I write about processes of ruination, I avoid referring to the sites I work with as ruins, partly because this label would fix their identity, and what I am most interested in is how these identities can remain unfixed yet still productive.

In this book, I follow processes of material dissolution and disintegration, and I attempt to describe the ecological and chemical processes that produce the effects we recognize as ruination. Throughout, I try to assemble resources that would allow us to locate our stories in the movement of matter. In the sense that this is an aesthetic project, it draws on a model of aesthetics akin to Eagleton's "gaze and guts" or Edensor's "emergent aesthetics."[56] My critical lens focuses not (only) on the surface layers but on attending to the way we encounter and apprehend things as they come undone and are drawn into other orders, other systems. This book is about locating the threshold, the point to which entropic process is allowed to run. It also asks what it would take to cross that threshold, to countenance finitude, complete dis-integration, and reclamation into other forms. In sympathy with Mark Jackson, I try to imagine how we might open ourselves to "decay . . . as an ontological ground for a post-humanist ethics."[57]

One of the things that I've come to realize is that receptivity to the kind of experimentation I'm proposing in this book will vary depending on the scale of the objects under consideration. My ideas originated with work in the fine grain of matter, where things could be seen to be passing from one state to another but where the overall integrity of the surrounding environment was not challenged. I have found it more difficult to attempt to tell stories about the unraveling of bigger things—built structures, like buildings and harbors.[58] Our minds have a tendency to consolidate these things as cultural objects, and it takes an extra effort to see them as provisional gatherings of matter, on their way to becoming something else. I needed to train myself to see both the form of the structure and the substance that it was made of, and to learn how to trace the web of relations that extended out from that substance.

Architectural theorists are fond of stressing the unfinished qualities of architecture and the ways in which buildings' lives are extended through acts of alteration, amendment, destruction, and wear.[59] Moshen Mostafavi and David Leatherbarrow write of building weathering as a "form of completion" and ask whether "it is possible that weathering is not only a problem to be solved, or a fact to be neglected, but is an inevitable occurrence to be recognised and made use of in the uncertainties of its manifestation."[60] In the sites I discuss in this book, the unfinished extends to the point of unmaking; even in states of near collapse, however, ruination does not signal the "absolute annihilation of building and organisation" but instead opens out into radically "different forms of organisation and organising."[61] Weathering and ruination can be understood as a form of self-excavation through which a structure gradually discloses its internal properties and material constituents.[62]

We are accustomed to thinking about buildings as whole and complete the moment their construction ceases, and preservation practice is largely oriented toward recovering this moment of wholeness and unity. If we accept that processes of aging and decay can be additive as well as destructive, then some form of temporal reorientation must take place as well. Almost all of the terms that are used to describe attitudes of care, toward both cultural artifacts and natural environments, assume the desirability of a return to a prior state: restoration, conservation, preservation, reconstruction. There are some more neutral terms in circulation, such as stabilization or consolidation, but for the most part, the gaze must snap backward to find its point of reference. In real terms, however, the people responsible for caring for both natural and cultural heritage often manage not recuperation but change, working with remnant ecologies and materials to produce conditions that draw on past precedents but move forward into new forms. We lack an appropriate language to describe this future-oriented practice, and reversion to the available terms often requires us to make excuses for invention and transformation rather than accepting it as a necessary condition. There are signs of a shift taking place in ecological circles, with an increasing acceptance of novel ecosystems and a departure from the attempted recovery of historic conditions to embrace the emergence of new trajectories.[63] In relation to cultured materials, the concept of adaptive reuse introduces a future orientation into heritage practice, but it stops short of countenancing uses by other-than-human organisms and agencies. Daniela Sandler's coining of the term "counterpreservation" to describe the deliberate cultivation of decay and decrepitude in reunified Berlin comes close, but a fully realized entropic heritage practice would require more sustained attention to the organisms and entities with which we share our world.[64]

I am fully aware that foregrounding entropic process in

our inherited structures and artifacts may be ultimately un-
workable in practice. Rather than locating my argument in
a theoretical space where I can make my point without interference from the clutter of the real, however, I want to
follow experimental practice to the point of failure. Luke
Introna writes, "The ethos of letting be is impossible—and
so it should be. . . . It is exactly this impossibility that leads us
to keep decisions open, to listen, to wait, and to reconsider
again our choices."[65] This book is full of stories about the
gap—the tense place between abandonment and attention.
I try to imagine what it might mean to dwell there, and let
things be unpredictable and permeable—not entirely known,
or owned, by us.

2

Memory's Ecologies

CURATING MUTABILITY IN MONTANA

If you are squeamish
Don't prod the beach rubble

Sappho, Fragment 84

M ANY OF THE IDEAS I EXPLORE in this book can be traced back to the time I spent poking about at a derelict homestead in Montana. The farm, settled with a homestead claim in 1889 by the Moon family, lay a few miles north of the city of Missoula, tucked into a swale in the bare Rocky Mountain foothills. For most of the twentieth century, another family, the Randolphs, ran a market garden and subsistence operation on the site, but by the 1990s the farm's productive days were long past. The youngest son died in 1995, leaving behind a complex of ramshackle sheds, barns, and dwellings, packed with domestic and agricultural debris. I came along in 1997 and began to work with the site's residual material culture, first as a volunteer curator and later as a research student working toward a doctorate in cultural geography. My excavations performed an ad hoc archaeology of the recent past in a place not yet old enough to be interesting to (most) archaeologists and too marginal and dilapidated to be a straightforward candidate for historic preservation.

As I worked in the homestead's abandoned structures, I often came on deposits of ambiguous matter that resembled

Georges Bataille's description of the "unstable, fetid and lukewarm substances where life ferments ignobly."[1] Maggots seethed in tin washtubs full of papery corn husks. Nests of bald baby mice writhed in bushel baskets. Technicolor mold consumed magazines and documents. Unpleasant odors escaped from the cracked lids of ancient preserve jars. Rodents, insects, and other organisms, long accustomed to being left alone, had colonized the excess matter. Packrat middens crowded attic corners with pyramids of shredded text and stolen spoons. Hoardings deposited by animals and humans mingled indistinguishably. I am not particularly squeamish, but the edge of revulsion was never far away. I worked close against the margin where the "procreative power of decay" sparks simultaneous—and contradictory—sensations of repugnance and attraction.[2] In my early excavations, the degraded material presented a problem that I could barely articulate, let alone resolve.

In her characterization of waste as a by-product of the creation of order, Mary Douglas comments on the threat posed by things that have been incompletely absorbed into the waste stream. "Rejected bits and pieces" that are recognizably "out of place," she observes, still have some identity because they can be traced back to their origins." "This is the stage at which they are dangerous," Douglas writes. "Their half-identity still clings to them and the clarity of the scene in which they obtrude is impaired by their presence."[3] Such obtrusions of clarity were common in my encounters with the homestead's artifacts: a bundle of paper furred with mold; a tangle of stained fabric and desiccated mouse carcasses; musty locks of human hair; a pair of badger paws tacked above a door lintel; tin cans cloaked with rust and cobwebs. These things were caught up in the processes of "pulverising, dissolving, and rotting," which would eventually render them unrecognizable.[4] Their disposal, however, remained unfinished.[5]

At base, the questions I raise in this chapter arose from my uncertainty about which items I should attempt to salvage from these deposits of comingled matter. In ruins, Tim Edensor comments, "processes of decay and the obscure agencies of intrusive humans and non-humans transform the familiar material world, changing the form and texture of objects, eroding their assigned functions and meanings, and blurring the boundaries between things."[6] As the curator of the site, I had responsibility for recovering items of value from this inauspicious mess so they could be enlisted for projects of cultural remembrance. I soon realized, however, that the things I was most drawn to didn't necessarily lend themselves to recovery. "It is unpleasant to poke about in the refuse to try to recover anything, for this revives identity," Douglas observes (echoing Sappho's "if you are squeamish" caution). Douglas goes on to suggest that such materials produce "ambiguous perceptions" that trouble the order of things.[7]

Conventional strategies for artifact conservation and heritage preservation neutralize these ambiguous perceptions through judgments that render materials into distinct categories of *artifact* and *waste*. In this place, however, such an approach would have led to the disposal of all but the most durable and discrete items. This is, in fact, what almost happened. The state university's lead archivist took one look at the massed clutter and reached for a black plastic garbage bag. The curator from the local historical museum refused to touch the homestead's documents and artifacts for fear of spreading their mold to her collection. The degraded condition of the materials mediated against their inclusion in public collections and archives. The homestead's materiality required a particular kind of attention to make sense of it, one that attempted not to defuse sensations of ambiguity and aversion but to work with them.

I begin this chapter with a discussion of the way certain deposits of material open up breaches in the categories we use to order the world and to structure our attempts at remembering the past. A section on memory work and heritage follows, with a proposal for a mode of remembrance that might accommodate these shifty materialities and yield to the collaborative energies of other agencies. I continue with a rumination on how such approaches trouble the authority of the curator, then move on to an experiment in collaborative curation. The chapter is primarily concerned with problems of interpretation: how can we work with these slippery things without eliding their ambiguity?[8] Peter Sloterdijk writes of the need for people who can work in a spirit of "liberating negativism," pushing past their nausea to confront material too unpleasant for others to contemplate."[9] The threshold of discomfort and aversion, Sloterdijk suggests, can also be a threshold to other ways of knowing.

I made a curious discovery one morning while picking at the debris in the homestead's old creamery shed, which had long since been given over as storage for miscellaneous matter. Against the shed's back wall, under a long bench, behind a heap of baling twine and feed sacks, sat a dingy wooden box, roughly two feet wide by four feet long. I pulled off a covering piece of corrugated tin to disclose a grayish mass of fiber and fragments that filled the chest up to its rim. Then I noticed what appeared to be a leather book cover, and another. A collection of battered volumes nestled among the litter. Leaning closer, I saw that scraps of torn paper made up part of the box's gray matter. I could make out a few words here and there: "shadowed," "show," "here," "start," "Christ." The printed paper mingled with mouse droppings, cottony fluff, plum pits, and dried leaves. Tiny gnaw marks showed along the spines of the books. I opened one mottled text, *Bulwer's Work*, to a chapter on "The Last Days of Pompeii" and read a passage about the inhabitants of that ill-fated town.

An Encyclopaedia of Practical Information occupied pride of place in the top center of the box. The chunky reference text seemed to be intact, save for a small insect borehole in the upper right corner of the first page above the publication date (1888). I carefully turned the brittle sheets to page 209, where I found a table on the "Speed of Railroad Locomotion." Page 308 detailed cures for foot rot in sheep; page 427 offered a legal template for a deed with warranty; and page 608 informed me that "Ecuador lies on the equator in South America, and is a republic." The borehole tracked my progress through the brittle paper. At page 791, a table recording the population of world cities (Osaka, Japan, 530,885; Ooroomtsee, Turkestan, 150,000), I had to stop, lest I crack the book's stiff spine. Below, the pages disappeared into the litter of seeds and scraps, the single insect hole still tunneling down into the unknown.

Faced with a decision about what to do with this mess, I balked. The conservator in me said I should just pull the remaining books out of the box, brush off the worst of the offending matter, and display them to the public as a damaged but valuable record of obsolete knowledge. Another instinct told me to leave the mice to their own devices and write off the contents of the box as lost to rodent infestation. I could understand the mess as the residue of a system of human memory storage or as an impressive display of animal adaptation to available resources. It was difficult to hold both of these interpretations in my head at once, though. I had stumbled on a rearrangement of matter that mixed up the categories I used to understand the world. It presented itself as a problem to be solved with action—putting things in their place. But what I found myself wanting to do most, after I recovered from my initial surprise, was to take what was there and think about how it got there. I wanted to follow the bookworm on its path through the encyclopedia.

In the box nest, I had come up against a moment of ambiguous perception in which my interest was torn between two apparently contradictory interpretive options. To borrow a turn of phrase from environmental archaeology, I found myself with a decision to make about whether I was looking at an artifact—a relic of human manipulation of the material world—or an ecofact—a relic of other-than-human engagements with matter, climate, weather, and biology.[10] Cultural matter had taken on an explicitly ecological function. To see what was happening required a kind of double vision, attuned to uncertain resonances and ambivalent taxonomies. "Thinking about natural history and human history is like looking at one of those trick drawings," writes Rebecca Solnit, "a wineglass that becomes a pair of kissing profiles. It's hard to see them both at the same time."[11]

If you're only attuned to see the wineglass—the evidence of explicitly human activity—then the onset of decay and entropic undoing may look only like destruction, an erasure of memory and history. Paying attention to one aspect of the object's existence deflects attention from another. But if we can hold the wineglass and the kiss in mind concurrently, decay reveals itself not (only) as erasure but as a process that can be generative of a different kind of knowledge.[12] The book-box nest required an interpretive frame that would let its contents maintain simultaneous identities as books and as stores of raw material for rodent homemaking. Michael

Taussig touches on a similar theme in an essay on the peculiar character of bogs and swamps. He muses on the ways
boggy, rotting places expose "the suspension between life and
death," flitting "between a miraculous preservation and an always there of immanent decay."[13] Taussig acknowledges how
difficult it can be to encounter amalgamated deposits of cultural and biological memory in these places: "What you have
to do is hold contrary states in mind and allow the miasma to
exude," he writes.[14] Taussig's advice seems promising, but
how exactly do we go about letting the miasma exude? This is
not a particularly easy thing to do, especially when curatorial
work assumes a certain responsibility for stabilizing things
in frames of reference that make them accessible to those
who come along afterward. I soon came to realize, however,
that the drive toward stabilizing the thing was part of the
problem.

In the past few decades, theoretical approaches that locate the identity of an object in its fixed material form have
given way to more complex notions of object identification
as a mutable and contingent process.[15] Most recently, work
has focused attention on the way that objects themselves can
be understood as "processual events," continually formed
and transformed by their movement through a field of social and physical relations.[16] There remains in museum and
material culture studies, however, a pervasive assumption
that the meaning and significance of an artifact can best be
sustained by securing, so far as possible, its physical permanence (a theme we will return to in subsequent chapters). Yet
routines of daily life often depend on the material modification of physical objects: people use things up and wear them
out, consume and combine.[17] Objects generate meaning not
just in their preservation and persistence but also in their
destruction and disposal.[18] Apparently destructive processes

play a crucial role in facilitating the circulation of material and ensuring the maintenance of social codes.

This is also true of objects transformed or disfigured by ecological processes of disintegration and regeneration. These things have social lives, but they have biological and chemical lives as well, which may only become legible when they begin to drop out of social circulation.[19] The disarticulation of the object may lead to the articulation of other histories and other geographies. An approach that understands the artifact as a process, rather than a stable entity with a durable physical form, is perhaps able to address some of the more ambiguous aspects of material presence (and disappearance). Tim Ingold, drawing on Heidegger, makes the case for "an ontology that assigns primacy to processes of formation as against their final products, and to flows and transformations of materials as against states of matter." In such an ontology, he suggests, even "ostensibly artificial structures," such as buildings, are not inert but are caught up in a continual exchange of materials; rain wears away paint, fungus decomposes timber, plants root in gutters, and human and nonhuman inhabitants come and go.[20] To become attuned to processes of formation requires cultivation of an inverted perception that resists the urge to settle the identity of the things we encounter and instead remains open to their continual material becoming. The book-box nest was neither artifact nor ecofact but both, a dynamic entity entangled in both cultural and natural processes, part of an "admixture of waste and life, of decadence and vitality."[21] Of course, in order to think this way it is necessary to resist (or at least defer) the urge to "save" the artifact. Interpretation requires a willingness to let the processes run and to pay attention to what happens on the way.

Though this might seem willfully destructive to those

who locate the memorial potency of the object in its discrete physical form, I want to suggest that a different kind of remembrance becomes possible when one engages in this kind of work. Others have drawn critical political and aesthetic insights from engagements with degraded and fragmented things.[22] Although these themes weave through the fringes of the analysis I put forth here, this chapter is not directly concerned with this body of work. I turn instead to a discussion of how the homestead's mutable artifacts allowed for an exploration of the blurry no-man's-land at the border between our categories of "natural" and "cultural" matter. It is here, where what we call *human* unravels into what we call *other*, that the ambiguous perceptions seemed to lie most thickly and promise most fully.

Edward Casey writes, "Everything belongs to some matrix of memory, even if it is a matrix which is remote from human concerns and interests."[23] In the third edition of the Collins English dictionary (1991), the ninth (and final) definition for the word "memory" reads, "The ability of a material, etc., to return to a former state after a constraint has been removed." The matter that makes up the homestead's structures and features exhibits just this kind of tracking backward, as well as a dynamic evolution into other states. Human labor introduced temporary arrangements—clear window glass, milled lumber, tempered fence wire. But these arrangements are unstable. Century-old glass develops cloudy irregularities in its gradual recrystallization. Faded scraps of newspaper mingle with desiccated leaves. Lichen grows on a standing building, a symbiotic association of fungus and algae breaking down milled clapboards to make them available for recycling into new growth. A lump of soft coal, pulled from the nearby mine seventy years ago, recalls the organic matter of a twenty-five-million-year-old forest. The homestead, like the abandoned Welsh farms described by Mike Pearson and Michael Shanks, is a place where "the very processes of the archaeological are apparent: moldering, rotting, disintegrating, decomposing, putrefying, falling to pieces."[24] The formation processes that create the archaeological record are here just getting under way.

It is exactly these processes of moldering and disintegration that most conservation practices work to forestall. In conventional terms, in order for the object to function as a bearer of cultural memory, it must be protected in perpetuity. Acts of counting, sorting, stacking, storing, and inventory convert things from the category of *stuff* to the status of museum object. As the curator at a historic Montana ranch managed by the National Parks Service commented to me, "If it's museum property it needs to be taken care of and preserved forever—that's kind of the responsibility of it being in that category."[25] (One of the items in her care was a fused mass of iron nails, glued together by extreme oxidation and recovered from a river dump site. It is stored in a climate-controlled facility on a padded shelf fitted with earthquake restraining straps.) Conservation technologies slow or halt physical decay, while interpretive strategies present the objects as symbolic remainders from a static past. Ephemeral things, decontextualized and cataloged, acquire a "socially produced durability" in carefully monitored environments.[26] Objects are stored in humidity- and temperature-regulated, rodent- and pest-proofed storage areas. Special paint protects artifacts from damaging ultraviolet rays; chest freezers decontaminate cushions and clothing of lingering mold and microbes. Arrested decay—the preservation policy applied to buildings when one wishes to maintain their structural integrity yet preserve their ruined appearance—also works at the scale of individual objects. Most practices oriented toward caring for the material past take great pains to ensure that the physical and biological processes that provide tangible evidence of the passage of time have been neutralized.[27] The memory encapsulated in the structures and artifacts at places like the curator's historic ranch references a resolutely human history, and any loss of physical integrity is seen as a loss of memorial efficacy—an incremental forgetting. But the state of affairs is, of course, more complicated than it appears to be. Strategies to arrest decay always destroy some cultural traces, even as they preserve others. And decay itself may clear a path for certain kinds of remembrance despite (because of?) its destructive energies.

A thicket of box elder trees crowds the fence line at the bottom of the homestead's decadent orchard. Given their girth and height, the trees appear to have seeded within the last half century. Long before then, the area along the fence accumulated an assortment of farm implements and stockpiled materials: a spike-toothed harrow, a stack of salvaged boxcar siding, a grain binder. Unneeded objects came to rest in the widening shade of the weedy trees, and no one paid them much attention. Eventually the trees began to draw the snarl of iron and steel into their generous vegetal embrace. The edge of a studded wheel fused into gray bark; a branch thickened and lifted over the binder's mass, carrying with it, and gradually consuming, a loose length of chain; roots twined around steel tines. The binder—designed to cut, gather, and fasten sheaves of grain—became bound in place. Pale lichen encrusted the driving chains that wound around the body of the machine. One of the binder's molded iron handles now protrudes from a slim trunk, as if to invite an adjustment of the systems of multiplying cell and running sap. The hybrid tree–machine works away at a perennial chore,

binding iron and cellulose, mineral and vegetable. The implement appears to be beyond saving, too broken down and
biodegraded for recuperation through conventional heritage
treatments. If you start to think about the decomposition of
the binder in another way, however, it is possible to see the
ongoing intervention of the trees and the soil as productive
of other resources for recalling the past in this place. An example from farther afield might help explain what I mean
by this.

Susanne Küchler's work in Papau New Guinea has documented the construction of *malanggan*, monuments to the
dead. Mourners construct these assemblages of wood or
woven vines and decorate the surface with carvings of animals, birds, shells, and human figures. The perishable monument is placed over a human grave as a marker. After a
certain amount of time has passed (when the human soul
is understood to have escaped the body), the *malanggan* are
taken from the graves and set in a location (often near the
sea), where they are left to rot. Once the *malanggan* have de-
composed, the residual matter is gathered to fertilize local
gardens. Küchler describes how this vital memorial tradition
turns "the finality of death to a process of eternal return."
The mode of remembrance practiced in the *malanggan* ritual,
Küchler argues, does not require a physical object for its operation but draws instead on the gradual erosion of this physical presence, the "mental resource created from the object's
disappearance."[28]

Küchler frames her analysis of this memorial practice
through a discussion of its antimaterialist qualities. What
strikes me, however, is not the rejection of materiality per
se but the embrace of the mutable character of material
presence, the transformative powers of decay and revital-
ization. Küchler (drawing on Walter Benjamin) asserts that

the "ephemeral commemorative artefact" might "instigate a process of remembering directed not to any particular vision of past or future, but which repeats itself many times over in point-like, momentary . . . awakening of the past in the present."[29] Cultural remembering proceeds not through reflection on a static memorial remnant but through a process that slowly pulls the remnant into other ecologies and expressions of value, accommodating simultaneous resonances of death and rebirth, loss and renewal.

I wonder if it is possible to approach the grain binder as a *malanggan* of the American West, which also releases its meaning in decay. An artifact of technological innovation sinks into the dark loam under the box elder trees and recalls its origins in veins of ore under the dark earth. The ruined machine sparks reflections on once robust economies, the changing markets and consolidations that precipitated the transformation of the American West's agricultural landscapes (and the gradual obsolescence of diversified small farms like the homestead). Raw material returns to the earth or is seized into the lignin and cellulose of a tree—the tree itself an import from another part of the continent, brought to the West to domesticate unfamiliar places. Now the weedy trees signal the inexorable "rewilding" of places that are left to their own ecological devices.[30] These suggestive interpretive resources would not be available if the binder were to be sawn from the tree, repaired and restored, and set alongside other mechanical agricultural dinosaurs. (Such a salvage effort is probably impracticable, at any rate, and would most likely lead to the destruction of both tree and implement.) The binder suggests a mode of remembrance that is erratic and ephemeral, twined around the past and reaching imperceptibly into what is yet to come. The trees participate in the production of cultural memory as "an activity occurring in

the present, in which the past is continually modified and re-
described even as it continues to shape the future."[31] Memory
is based on chance and imagination as much as evidence and
explanation; the forgetting brought on by decay allows for
a different form of recollection. Such recollection fosters an
acknowledgment of agencies usually excluded from the work
of interpretation.

The farm's root cellar—a cavernous space with crumbling
earth walls and a pervasive scent of sour rot—disclosed an
archive of historic documents stashed in its dim corners and
dusty crates, each item spectacularly degraded in its own way.
One excavation turned up a coiled map with a deeply nibbled
edge. When I unrolled the coil, the chewed section unfurled
to reveal a repeated pattern in an ornate fringe, like a paper
doll cutting. The fringe ate into a gridded territory that represented the United States Forest Service management districts
just west of the Missoula Valley. Insects had intervened to
assert the materiality of the map, and in doing so, they offered their own oblique commentary on human intervention
in regional ecologies. The forests in the physical territory de-
picted by the disfigured paper map had suffered from decades
of poor management and fire suppression, which made them
vulnerable to the depredations of other organisms. Over the
last few decades, an infestation of destructive bark beetles
has killed many of the trees represented by the map's green
patches. The destruction on the root cellar's map could be
read as a metonym for the destruction of the surrounding
forest. The disarticulation of a cultural artifact allowed for
the articulation of other histories about invertebrate biographies and appetites. This speculative act of allegorical disclosure worked through a principle by which "objects have to fall

into desuetude at one level in order to come more fully into their own at another."[32]

Other documents showed equally impressive evidence of insect and rodent intervention. In their degraded condition, these documents carried an unusual charge. I had come up against an absence in the record, but an absence that seemed to open a window in the wall that usually keeps cultural analysis separate from the investigation of ecological process. It required some imagination to work past the initial awareness of missing information, but once this had been overcome, I could see the emerging shape of an engagement with the past that drew part of its force from absence and fragmentation.[33] Christopher Woodward, in his observation of the creative resources that people generate when confronted with ruins and remnants, identifies a sympathetic association between structural incompletion and imaginative engagement.[34] Degraded artifacts can contribute to alternative interpretive possibilities even as they remain caught up in dynamic processes of decay and disarticulation. The autonomous exercise of human control gives way to a more dispersed sharing of the practices of material editing and curation.

Miles Ogborn, in a short essay on the ecology of archives, writes about how archives and their contents, which arise out of a patently cultural desire to preserve the human past, are also amalgams of animal skin and wood pulp, chemical compounds and organic substances. The elements that make up the archive are open to breaches and interventions—from heat, light, moisture, mold, insects, rodents. Ogborn writes, "The storehouses of memory, the central cortices of social formations of print and the written world, are ecologies where the materials of remembrance are living, dying, and being devoured."[35] The "nature of cultural memory" becomes apparent in the gradual consumption and erosion of evidence and images.

I found one of my favorite examples of these cultural ecologies at work in a battered copy of *National Geographic* magazine, which had been stored with others of its kind in a set of cranberry crate shelves in the farmhouse kitchen. In the forty-year interval between human habitation of the dwelling and my intervention, hungry mold and rodents had consumed the glossy pages. When I went to open the magazine, this particular copy peeled apart reluctantly to reveal a patchy scene brushed with delicate pink. The mold had eaten away an image of a mountain town to expose a few bars of music, an area of green, shards of unintelligible text. There was a curious loveliness to the transformed scene—mountains and music and mold in a montage of indeterminate effect. The cultural spore of mass printed matter was caught up in the fungal ecologies of decay, its authority an impartial documentation of a world *out there* undermined by the microscopic imperatives of a world *in here*.

The homestead's entangled artifacts also worked to remember the past in place on another register. The farm's shacks and sheds were packed with collections of miscellaneous material: sacks stuffed with feathers and leaves, bushel baskets of wool and fiber, neat stacks of twigs, jars of seed and sand. It was not always clear to me how these collections had been assembled, and they often troubled the distinction I tried to draw between animal and human labor. The homestead's tack shed contained a few tin cans filled with fruit stones—rough pits from the wild plums that grow in the gully, and the hard seeds of the orchard pie cherries. Each of the stones was neatly scraped down to its woody center and marked with a tidy chewed hole, through which an animal had extracted the edible core. In the farmhouse pantry, I found packets and tins of saved garden seeds assembled through a sympathetic impulse, put by as insurance against anticipated scarcity. The root cellar's dusty shelves held dozens of cloudy jars of preserved cherries, rhubarb, and tomatoes. In a crate below the shelves, I found a 1937 postmarked envelope full of flower seeds and a twist of catalog paper around a handful of white snail shells. Nearby a stained pillow hung from a nail, the bottom eaten through to let out a slow leak of feathers.

An odd affinity seemed to hang over these accumulations; intertwined memories of seasonal harvest and hoarding seeped out of the jars and tins and sacks. The collections allowed me to see the human activity that went into constructing and provisioning the homestead as just another layer of habitation, and I began to appreciate how both humans and other organisms draw in the raw material of their world and animate it through their complex practices of dwelling and making.[36] The place presented a mingled material record deposited by several different species of extended organisms, and the memory in these accretions of matter spoke to decades of cohabitation, of entangled lives and habits. People inhabit places through accumulations of books and tools and clothes and seeds; mice inhabit places through accumulations of pits and leaves and bones, as well as their reworkings of the matter people leave behind.

The finest grain of the (elusive) boundary between animal and human habitation lay in the dust. As one of my first curatorial acts at the homestead, I sorted the contents of the long-abandoned kitchen junk drawers. After I had removed and set aside the recognizable household objects (lamp wicks, citrus wrappers, clothespins, string balls, thimbles, spools, pencil stubs, bottle stoppers, shoelaces, keys, washers, heels, hooks), a layer of fine-grained detritus remained at the bottom of the drawer. On close inspection, I was able to identify bits of mouse droppings, rubber shreds, wood splinters, paper, lint, wire, insect wings, plant stems, seeds, human hairs. An even finer grain of residue underlay these legible fragments, a slightly greasy amalgam of human skin, tiny fibers, crumbled deposits of mineral and animal origin. I remember feeling dizzy while I examined these leavings, sharply aware that I had reached the base level of materiality, the place where human artifacts blended imperceptibly into mass of worldly matter.[37] As Phil Dunham points out, encounters with dust raise questions about "what (if anything) is consistent or whole about our bodies, and where (and indeed whether) a line can meaningfully be drawn between the human and nonhuman worlds."[38] These encounters, though sometimes unpleasant, served as a powerful reminder of my own entangling with these borderline materials and their active ecologies.

My early decision to let the dust and the detritus into my interpretive frame was not without its risks. As I began to allow myself to yield to these messy remains, I realized that in order to meet them (so to speak) on their own terms, I had to accept that the outcome of the situations I found myself in was not entirely in my hands. "Expelling and discarding is more than biological necessity—it is fundamental to the ordering of the self," observe Gay Hawkins and Stephen Muecke.[39] In choosing not to discard materials that would, in other circumstances, have been quickly consigned to the rubbish bin, I also opened myself up to influences that unsettled my sense of curatorial authority and allowed the material to "act back" on me in unexpected ways.[40] Once I'd deviated from established standards of valuation and significance, the sheer excess of eligible material mocked my attempts at recovery and rationalization. Sometimes I found myself pushing back against the chaos to assert some kind of (usually ineffectual) order. But I also experimented a bit with strategies that took on the forces of decay and deterioration as allies rather than adversaries.

One day I came across an overstuffed bushel basket in the homestead's harness shed. I pulled out the top layer of stained clothes to disclose a stew of paper, fabric, and animal leavings. I tipped the whole thing on the grass, and in the scatter I identified scraps of printed matter mixed in with a mass of pits and seeds, woolly fiber and feathers, long johns and holey socks, a 1928 licence plate, and a few delicate mouse spines. I had come across similarly scrambled deposits countless times in my excavations, and I usually gave in to the impulse to discard or burn all but the most discrete items—in this instance, only the licence plate presented itself as immediately eligible for salvation. This time, however, something about the mess drew me in, and I began to pick shards of text out of the other litter. Later, I took some liberties and drafted a poem from the salvaged fragments:

> the camera
> may
> record
> odd
> invented
> museums
> placed at
> glare horizon
> your service
> makes
> value
> almost anywhere

cardboard box
on the wall
behind a picture
square-mile ice
parts will
have a
numerous
synchronic
handiwork
that of
invention in
minimum
delight

I like to think that the mice and I share authorship for this work—with some credit due as well to the authors of the articles in the shredded magazines (which I have tentatively identified as an amalgam of 1940s *Popular Mechanics* and Seventh Day Adventist religious tracts). I suppose I should also mention Tristan Tzara, whose dadaist poem instructions run like so:

Take a newspaper
Take some scissors
Choose from this paper an article of the length you want to
    make your poem
Cut out the article
Next carefully cut out each of the words that makes up this
    article and put them all in a bag
Shake gently
Next take out each cutting one after the other
Copy conscientiously in the order in which they left the bag
The poem will resemble you.[41]

The salvage poem perhaps says more about my intervention in the homestead's sedimented histories than it does about the actual substance of those histories. But I include it here to suggest the terrain that might be explored by an interpretive practice willing to engage in serious play with artifacts that might otherwise be overlooked entirely.

This experimental engagement with some of Douglas's dangerous things allows other "sensible forms" to work alongside the curator in the generation of research materials.[42] In this instance, an act of "synchronic handiwork" takes up the raw material of the past and works it into a missive that speaks both to that past and to the lived present. The method celebrates the artifact's status as a temporary arrangement of matter, always on its way to being something else. At Hyde Park Barracks, near Sydney, Australia, rats collected the ephemera of decades of daily life in their nests between the floorboards. When conservationists and curators discovered these hoards, they decided to create a display to celebrate the findings. "Rats are honoured at this site as the minions of history," writes Barbara Kirshenblatt-Gimblett.[43] I occasionally attempted a similar strategy at the homestead, opening up my curatorial activities to the intervention of other organisms. Such strategies may generate interpretive ambiguities, but they also open up different ways of ordering the world and allow us to work past an entirely negative reading of material dislocation and dissociation.

The interpretive approaches I sketch out in this chapter—observed decay, ephemeral commemoration, collaborative curation—are presented in a speculative spirit. I recognize that the active cultivation of material dissolution and disappearance I advocate here would be impossible to achieve in most heritage contexts. This kind of approach is most appropriate, perhaps, for work with materials that lie at the fringes of conservation practice, or with things and places that are held in a state of limbo before more formal arrangements around preservation and public access take hold. It is possible, however, to imagine how established museums and heritage sites might begin to introduce a focus on material process (and a whiff of miasma) into modes of interpretation that tend to come down heavily on the side of stasis and preservation. Even a subtle shift in interpretive focus would require some attempt to hold those contrary states in mind—to accept that the artifact is not a discrete entity but a material form bound into continual cycles of articulation and disarticulation. When I was able to pull it off, this altered perspective allowed me to see things that otherwise would have been invisible to me simply because I lacked discursive frames in which to fit them. Interpretation, in this sense, constituted otherwise unconstituted matter.[44] I was able to read the messages on a wall of tattered newspaper scattered with box elder seed, the occluded histories in a rodent nest, and to include in my sense making the stories of other-than-human inhabitants usually entered only into the margins or consigned to their own separate texts. Instead of asking artifacts to speak to a singular (human) past, such a method works with an ecology of memory—things decay and disappear, reform and regenerate, shift back and forth between different states, always teetering on the edge of intelligibility.

Over a decade has passed since I concluded my curatorial work at the homestead, although after I received my doctorate I returned for a couple of years as a caretaker, and I was living there when my son was born in 2007. The community organization responsible for managing the homestead recently drafted a new strategic plan to guide their work, an update of the plan we had adopted in 2003. The plan notes that there remains "an ongoing tension between the Homestead's preservation and re-tooling and its natural decomposition. Such incongruity is thematic here. What should be saved? . . . What should be allowed to subside into the earth at its own pace?"[45] In practice, very few of the homestead's structures have been allowed to "subside," given the pressure to maintain safe public access and to stabilize as many of the historic structures as possible. One small, ordinary shed escapes mention in the new plan. Weathered and wracked, it has stood for years now in a state of near collapse, appearing to gradually twist itself into the earth as its angle of repose deepens. It may be the sacrificial subject, the test case for a light-touch entropic heritage practice. Meanwhile, all of the artifacts that I recovered during my time at the homestead— including jars of rodent-stored seed and the elaborately edited documents—are housed in the reconstructed homestead claim cabin, in cardboard boxes on steel shelves. The mice have moved back in and are rearranging the inventory.

3

When Story Meets the Storm

UNSAFE HARBOR

> Time present and time past
> Are both perhaps present in time future,
> And time future contained in time past.
>
> *T. S. Eliot, "Four Quartets: Burnt Norton"*

W E MOVED FROM MONTANA TO CORNWALL in December 2007, and our arrival coincided with the arrival of the season's first Atlantic storms. A day or two after we touched down, we drove over to Mullion Cove, on the westernmost edge of the Lizard Peninsula. We followed the road down from the village into a narrow valley and heeded the "no parking beyond this point" signs to pay our 40p and continue on foot. A few hundred yards down the road, the valley suddenly opened into a steep-sided cove. In front of us and to our right, an L-shaped stone breakwater reached out from ragged cliffs, like an arm crooked to protect an exposed face. A shorter arm extended out on the left, toward the tip of the other, leaving a gap between as the entrance to the small harbor. A scarf of spray rose above the breakwater with each swell, and the gale churned the sea in the harbor enclosure. My husband walked out across the slick stone and climbed up on the parapet, disregarding the warning sign: "Caution. Waves sweep over the breakwater in heavy seas. Keep well back."

I'd learned about Mullion while still in Montana, from an

article that appeared in *Preservation* magazine in 2006.[1] The article led with a question: "Storms, floods, and rising tides threaten many historic structures in the United Kingdom. So why is Britain's National Trust so willing to let them go?" The author described the "pragmatic, controversial, arguably fatalistic" policy that the National Trust had recently adopted in deciding to adapt to, rather than resist, processes of coastal change—even where, in some places, this would lead to the eventual loss of buildings and structures. The policy had been communicated in a 2005 report entitled *Shifting Shores: Living with a Changing Coastline* and related management principles, one of which bluntly stated, "Valued cultural features in the coastal zone will be conserved and enhanced as far as practicable, whilst not necessarily seeking to protect them indefinitely."[2]

The author of the *Preservation* article singled out Mullion Harbour as a site that at the time was on the front line of the Trust's coastal change policy. A benevolent landowner built the harbor in the 1890s, the National Trust acquired it in 1945, and English Heritage listed it as a structure of "special architectural or historic interest" in 1984. Since the acquisition, the harbor had sustained frequent, often severe, damage from storms and wave surges, and in 2006, the Trust, in consultation with the community, made a difficult decision:

> Once maintenance and repair is no longer deemed viable, the managed retreat phase will begin. In this phase, regular maintenance of the breakwaters will cease and the Trust will systematically remove the breakwaters whilst consolidating the inner harbour walls. However, the timescale for the move to managed retreat cannot be pinpointed, as it depends on when and how the ultimate extreme storm event or series of events occur.[3]

Mullion, it seemed, presented an opportunity to scale up the ideas I had been developing through my work at the homestead, and to think about memory and mutability in relation to an object much more substantial than my map scraps and mouse nests. Here was a whole harbor, flagged for finitude, and an organization, whose motto promised to protect "special places, forever, for everyone," that was confronted with the daunting task of explaining why, in this instance, it would only be "for the time being." The withdrawal of the promise of perpetual protection at Mullion was linked to a broader shift in coastal management within the United Kingdom in response to predictions of increased storm intensity and sea level rise. While for decades the policy had been to "hold the line" against coastal change with hard defenses, as the probable scale and scope of global climate change became apparent, there was an increasing openness to policy options described simply as "do nothing," or "no active intervention."[4] In 2007 it remained to be seen how these policies would be put into practice in specific places.

I visited Mullion often over the next several years, becoming familiar with the place and talking to the people who managed it about how they planned to navigate the transition ahead. Life in the cove went on. The Trust repaired the damage caused by each annual winter storm season. The harbor persisted, seemingly unmoved by its deferred death sentence. As time passed, I realized that the conundrum facing the National Trust at Mullion was as much about how to tell the story of the harbor and its future as it was about how to manage the eventual demise of the physical structure. I began to think about how the narratives used to describe places like Mullion tend to project long-term preservation indefinitely forward. When confronted with the impending transformation, or even disappearance, of such places, the usual narratives leave us with no other option than to see change as loss, and the withdrawal of care and maintenance as failure. I wondered whether it might it be possible to experiment with other ways of storying such transient landscapes, framing their histories around movement rather than stasis and drawing connections between past dynamism and future process. Eventually I assembled a reverse chronology that scrolled back to before the harbor's construction, into the cove's geological origins, looking to the harbor's fractured past to find resources for encountering its future unmaking. An article, described as an experiment in "anticipatory history," appeared in the *Journal of Historical Geography* in 2012, and I moved on to other projects.[5]

Then the storms came. In early 2014, a series of strong low-pressure weather systems rolled in off the Atlantic, building to a storm season more intense and incessant than any in living memory. As Cornwall's landmass checked the rolling progress of each successive swell, massive waves scoured the exposed coast and the structures along it. By the time the storms moved on, my carefully crafted future history of Mullion was broken, in that it could no longer do the work I had intended it to do. It had come up against the volatility and unpredictability of the storms, but it had also been exposed to other stories, assembled by people who had their own way of making sense of the past and the future in this place. This chapter is in one sense about the limits to narrative. If we are looking for ways to let go gracefully, story can provide some solace. But the words that we stack up and ask to do our work are not always enough; they can be unsettled and undermined as surely as a harbor wall. How do we make sense of the world, and our place in it, when language fails?

When I assembled the narrative about Mullion, one of the

things I was concerned to do was, as I stated then, "to open up an appreciation of the past not as static and settled, but as open and active."[6] I wound the clock backward to tell the story of the landscape's formation 300 million years ago and then I described how the harbor came to be assembled in the 1890s through a reworking of ancient geological matter— local serpentine and slightly less local granite. I chose specific historic moments of damage and near collapse to illustrate how the harbor was a product of continual rebuilding and reassembly, stabilized as much by its designation as an object of heritage as by physical acts of maintenance and repair. T. S. Eliot wrote in *Four Quartets*, "History is a pattern / Of timeless moments."[7] With my experiment in anticipatory history, I wanted to challenge this perceptual inertia and to give "time back to a timeless landscape."[8] I weave the retelling of Mullion in this chapter through with my rereading of Eliot's poem, letting his words nudge my thinking along and unsettle earlier certainties.

> Either you had no purpose
> Or the purpose is beyond the end you figured
> And is altered in fulfilment. There are other places
> Which are also the world's end, some at the sea jaws,
> Or over a dark lake, in a desert or a city—
> But this is the nearest, in place and time,
> Now and in England.[9]

In this chapter I return to my broken story to think about how it may have been "altered in fulfilment" by the events of the past winter. I stay close to the events as they unfolded and try to communicate a sense of how they affected me and the other people who experienced them. This new narration is not particularly tidy, and it doesn't have a clear ending. In the sense that I have a method, it is one that, as Michael Taussig suggests, uses "defective storytelling as a form of analysis,"[10] worrying over the details to share some of my perplexity about an unsettled season.

At the end of November 2013, I brought a visitor from California down to Mullion to show her the harbor and to see it for myself after months of absence and preoccupation with other research projects. As we walked out on the harbor walls, I scanned for signs of weakness and noted the most recent round of repairs. On the western breakwater, we could see patches where the granite cobbles (also called setts) in the walkway had been recently replaced and repointed; the new grout between each block stood out, bright and unblemished. On the southern breakwater, a large section at the end had recently been patched with concrete rather than the original stone, and there was a visible crack along the outer edge. This particular section had a history of instability and collapse: in the early 1950s, the squared end nearest to the harbor's sea entrance had been set back and replaced with a sloping concrete face, in the hopes that would it would deflect the force of the waves more effectively. The experimental engineering eventually failed. In 1978, the section was reconstructed in stone to an approximate replica of the original design. As with Tim Edensor's study of the building stone of Manchester, "palimpsests of repair work . . . [had] been repeatedly superseded by new restoration methods, all enacted with the aim of resisting the tide of material destruction and entropy."[11]

I was lying on my belly trying to frame a photograph of some rebar staples on the inner coping stones when a man approached us. He made a quip about my awkward pose, and I scrambled up to face him. I don't remember how our conversation began, but he soon offered that he'd recently published a book on the history of Mullion Cove and the harbor.[12] Gesturing to Mullion Island off the coast, he told us that sometimes 400 ships at a time used to seek shelter along this stretch of coast in "Mullion Roads," waiting for the wind to shift so they could round Lizard Point. The piers were built to provide safe harbor for some of these ships, he explained, and to allow trade goods to be transported through Mullion village up the hill. He insisted that, contrary to the historical information about the site provided by the National Trust, the harbor wasn't built exclusively for the pilchard fishermen. I asked, cautiously, what he thought about the future. "They aren't telling the true history of the harbor," he said. "If they did then they would have to recognize how important it is, and maintain it properly." I mentioned that I had my own interest in Mullion's history; he took down my name and shared his own, Bob Felce.

After leaving the harbor, we picked up a copy of Bob's book at the art gallery at the edge of the village, and then I paid a visit to Alastair Cameron, the National Trust property manager responsible for Mullion, in his office at a nearby estate. Alastair mentioned that they were planning to review the harbor study in 2015, with the intention of establishing a "tipping point" for the transition to managed retreat. He admitted that he had hoped it would be obvious when the time came to let go. I suggested that perhaps, in a way, the surrender of the harbor to the sea had already begun with the most recent round of repairs, and that cumulative alterations might eventually create a patchwork structure, with concrete gradually edging out the original stone.[13] Alastair seemed to think that the transition would have more clarity than this, and reminded me that they were still committed to maintaining and repairing the structure for the time being. I asked if

I could be kept in the loop on any decisions they made about the harbor over the winter. Alastair agreed, but he also said that because their current management was essentially reactive, the Trust's level of involvement would depend on the kind of winter we had.

A couple of days later, an e-mail message arrived from Bob. In a slightly cool tone, he mentioned that he'd found my article about Mullion on the Web and was now aware that we saw matters quite differently. He implied that I had been involved in the 2005 deliberations about the fate of the harbor, in support of the National Trust's position. He queried my credentials and made reference to "geography in my day" (which involved a reading of *A Blueprint for Survival* as a student at Leeds). He also corrected a few errors of interpretation in my work—pointing out, for example, that the decay of the harbor during World War II had to be understood in the context of the shortage of labor during the war, with priority being given to the reconstruction of Britain's bombed cities. He restated his observation that the harbor had never "achieved its true position in Cornish History."[14]

At the bottom of his e-mail, Bob had pasted a link to his website, which turned out to be an extensive, sometimes exasperated, documentation of his take on significant moments in the harbor's past and present, with a strong editorial conviction that the harbor owners, the National Trust, had a moral and social duty to maintain and protect it.[15] The website reproduced passages in full from his published book, interspersed with notes about his visits to the harbor, progress reports on harbor repairs, excerpts from historical texts, and anecdotes about people, weather, fishing, and other miscellany. A subpage of the website—"Climate Change or Weather Patterns?"—set out to question the basis on which the harbor study decision had been made and expressed skepticism about predictions of increased storm frequency and intensity in coming decades. The site reproduced vivid archival accounts of intense and violent storms that had battered the coastline in the past, many of them associated with wrecked ships and lost lives.

I had come up against the counterstory to my anticipatory history. In Bob's narration, the events of the past were assembled as evidence not to prepare for harbor's future unmaking but to shore up the case for investment in its continued persistence. The book ended with a melancholy appeal: "The story of the Harbour and Cove has now been told. It is an iconic site, in a beautiful location, with a long history, a story worth telling, and which perhaps is deserved of preservation. Sadly, at the present time, this looks to be an unlikely outcome."[16] Bob and I each had offered a story to Mullion; our stories attempted not just to convey the history of the place as we understood it but also to effect a change in perspective and priorities. If the narrative I offered to Mullion in the first instance had skirted around its instrumental intent, Bob's counterstory flushed it out into the open, making it visible for what it was: an act of imagination that was oriented toward a specific, and selective, version of the future.[17]

In response to Bob's e-mail, I explained that I had not been involved in the Mullion study, as I'd moved to Cornwall the year after it was completed. (I later learned that Bob had also moved to Cornwall in 2007.) Both of us were trying to make sense of place through our narration of it—mine in an admittedly academic mode, Bob's more embedded in the thick of the place and community life there. His book was rich with material contributed by his new neighbors, while my story was thinner, more inclined to take the long view.

I wondered whether our two narratives canceled each other out or whether they were perhaps twinned in a kind of centrifugal teleological spin. Let go. Hold on.

> Time the destroyer is time the preserver . . .
> And the ragged rock in the restless waters,
> Waves wash over it, fogs conceal it;
> On a halcyon day it is merely a monument,
> In navigable weather it is always a seamark
> To lay a course by: but in the sombre season
> Or the sudden fury, is what it always was.[18]

The storms began on Christmas Eve and ended on Valentine's Day. At first there was almost a sense of festivity in the air as people went down to the sea to be close to the "sudden fury." It was only later that the tone turned darker and more anxious. On January 2, my husband and I went down to the harbor with our six-year-old son. The crack on the southern breakwater that I'd noticed in November had widened, and some of the new pointing around other blocks had washed away. A few hours before high tide, the waves were already casting an occasional wash over the breakwater, and my son ("I want to get wet!") tried to edge into the spray zone. A couple of hours later, an unsuspecting father raised his small child over the edge of the outer parapet to see the waves rolling in, and one of the waves came over the parapet to meet them. They were thrown onto the granite setts of the harbor wall in an angry wash of surf, and someone snapped a photograph. Within hours it was on Facebook, and within days it was on the cover of the British tabloids. The *Daily Mirror* headline read, "The Mad Dad: Engulfed by Wave as He Carries Tot on Sea Wall after Ignoring Warnings."[19]

During the next several days, the storms intensified. On

the evening of Friday, January 4, Bob posted on his website that the tide was higher than he'd ever seen it. The winds were driving the sea up the slipway and threatening the boats secured at the top. He cited the coincidence of another "perfect" storm that had occurred on January 4, 1867, when a series of wrecks in Mullion Roads claimed many lives. After the 1867 storm, public concern about the vulnerability of ships on the exposed coast led to the establishment of a lifeboat station in Mullion.[20] "Conditions were the same as they are now," wrote Bob, "and the storm has been repeated. . . . The tide continues to come in and out. Perhaps what goes around, comes around."[21] When the tide receded, it was clear that the crack on the southern breakwater had deepened and fractured further, and the wash cast up by the incoming waves appeared to be running through the structure rather than over it.

Storms continued to drive big swells against the coast. One day that week, I drove out at dusk to park on the headland above the harbor. Huge breaking waves cast skeins of spray over the top of Mullion Island, and those that weren't caught by the island rolled on to crest over the harbor walls. The next morning, the local radio station broadcast interviews with Justin, the National Trust ranger for Mullion, and Bob. There had been further damage to the southern breakwater, and the waves had loosened and launched some of the setts that faced the walkway on the western breakwater. The radio story took on an alarmist tone: "Should Mullion Harbour be allowed to slide into the sea?" the presenter asked. Bob came on air to say that the harbor was an "iconic part of the Cornish coast" and should be protected. Another local resident asserted, "This is heritage. . . . It should be maintained." Justin commented on the "incredible weather" and the "substantial damage" to the harbor. They would assess the damage in the spring, he said, but in the long term, "it's unsustainable," and the National Trust stood by their intention to eventually allow the cove to "return to its natural state."[22] I saw Justin later in the week, and he commented that the whole thing had been blown out of proportion. "There's no story," he said.

A couple of weeks later, Justin was called on to speak at the local parish council meeting about the damage to the harbor and the National Trust's response.[23] He reassured everyone at the meeting that the harbor was not being abandoned, and he mentioned that a salvage operation had begun to recover some of the 2,000 granite setts that had been dislodged by the waves and cast into the harbor. The minutes note that a member of the public raised a question about the long-term fate of the harbor; this person asked whether the structure's Grade II heritage listing would need to be revoked should the decision be made to proceed with managed retreat. Listed Building Consent would be required for any substantive work carried out on the harbor, Justin explained, and should the decision be made not to proceed with rebuilding, consent would need to be sought for this as well. He also noted, however, that damage appeared to be relatively minor, and the present intention was to proceed with a program of maintenance and repair.[24]

The next round of storms arrived in early February. On the night of February 4, the power cut out just as my family was sitting down to dinner at our house, ten miles from Mullion. We ate by candlelight as the radio propped in the window spooled out frantic news of trees down, waters rising, buses blown into fields, and train lines blocked or washed away completely. Winds across the county were clocked at over ninety miles per hour. Ten thousand homes were without power, and thirty-foot swells pounded the coast. Exposed towns and villages had been evacuated. The next morning, we heard Bob on the radio, reporting from Mullion: the waves were up at the top of the slipway again, the fishing boats were threatened, and the western breakwater was missing more setts and coping stones. Bob had talked to one fisherman who said, "We can't do anything more," and he had spoken to another fisherman's wife, her husband away, who was in tears.[25]

I canceled my morning seminar with an excuse about blocked roads and drove to Mullion with my husband. From our vantage on the cliff near the hotel, we could see the full fury of the storm battering the cove. Enormous waves pounded the harbor walls and visibly loosened and shifted individual stones. Their unrelenting impact was gradually dismantling the structure, piece by piece. Many more of the western breakwater paving setts had been dislodged, and each new swell appeared to nose another into the seething enclosure. We saw a section of the southern breakwater railing sway, then fall. The brittle sound of stone on steel reached us from hundreds of yards away, over the fierce howl of the wind. I noticed for the first time a heap of salvaged stone at the top of the slipway, covered with a flapping blue tarp. Down the coast in Porthleven that morning, the waves breached the inner harbor. No one could remember this ever happening before.

What force did my narrative have against this event? In the midst of the mayhem, my reverse chronology seemed flimsy and self-indulgent, a loose strand of seaweed tossed on the harbor wall and dragged off again just as quickly. I had spent years thinking about change and transience in this place, about how it could be navigated and negotiated. I had offered up a story as an antidote to loss. But now that the unraveling had begun, the story seemed starkly irrelevant to the lives of the people who lived in the place, and who were now watching it fall apart. Later that day I received an e-mail from Bob with the subject line, "Sad day."

The sadness was real, but what exactly was at risk of being lost? The physical presence of the harbor, yes, but also what it symbolized: both access to the sea and protection from it; the promise of safe harbor and a vantage point from which to observe the (episodically) unsafe waters beyond. The harbor was a vestigial link to now all but extinct livelihoods and an anchor for a small community's identity—a monument in the conventional sense, in that the object held the communal memory of the past in place. The harbor's disintegration risked the unmooring of this identity, although, paradoxically, this identity was also grounded in a memory of other storms that had battered the harbor and left damage in their wake. The people who lived in the cove took pride in weathering these storms, understanding in their bodies what it meant to "weather" as an active verb. Bob's e-mail reported that the county council had attempted to evacuate the cove residents during the storm, but the only people to leave were the temporary inhabitants of the holiday cottages. Everyone else decided to stay put.

The last big storm blew through on Valentine's Day. It caused more damage to the harbor walls and punched a hole in the roof of the net loft, the oldest building in the cove. As the weather finally settled, people in Mullion began to take stock of the damage and pick up the pieces. According to Bob's documentation on his website, a substantial section at the knuckle of the southern breakwater had collapsed, and over 6,000 paving setts had been lost from the western breakwater. Many of the massive granite coping stones that edged the inner walls had been tossed into the harbor as well. Local fishermen led the salvage effort with help from a few of their neighbors, residents from the village, and National Trust volunteers. Much of the heavy work was done using the dumper that usually helped launch the fishing boats. A digger was brought on site to recover the larger stones, but most of the material was moved by hand. The piles of salvaged setts at the top of the slipway around the capstan and the boats grew. Bob sent me an e-mail on February 16 that began, "A future on the edge."

By this point in the winter, the anxiety about the extended spell of unusually violent weather had saturated the media. It was as if people had suddenly woken up to the recognition that processes discussed abstractly and dispassionately for years—adaptation, resilience, managed retreat—were now both pressing and real. Debates broke out about whether the intensity of the storms could be attributed to climate change. The announcers on the local radio station asked the shadow environment minister, "Would you concede that there are places where people may not be able to live in fifty years?" She dodged the question. A local member of Parliament, queried about Mullion ("Should we let nature take it?"), adopted a reassuring line: "Where we can protect communities, we will."[26] Prime Minister David Cameron offered a blank check for storm repairs and free sandbags for all. When public debate finally began to countenance the possibility of letting go, the response was to retreat into defensive mode and to look away.

> The houses are all gone under the sea.
> The dancers are all gone under the hill.[27]

At Mullion, the National Trust was faced with determining whether the events of the past months constituted the "ultimate extreme storm event or series of events" required to

set in motion the transition to managed retreat. The con-
sensus, in the face of mounting public pressure, seemed to
be, "Not quite." A National Trust statement posted on February 24 alongside photographs of the salvage effort outlined
the situation:

> Despite the extreme pounding the harbour received during
> the series of storms, and a substantial amount of damage in
> the process, it is still in remarkably sound condition, par-
> ticularly the western (main) breakwater which, aside from
> some structural damage to the parapet wall, appears to
> have suffered mainly "cosmetic" damage. . . . The southern
> breakwater has suffered significantly worse damage. This
> breakwater was never built to the same standard as the
> main western breakwater, and consequently suffers greater
> storm damage.
>
> We're still awaiting the full report from the engineers,
> but subject to funding, (the estimated costs of these repairs
> is likely to be in excess of £1/4 million), we hope to start
> work on repairs to the western breakwater within the next
> few weeks. Any repairs to the southern breakwater are very
> dependent upon the results of the engineers' report and
> subject to Listed Building Consent. They will more than
> likely be undertaken using a combination of concrete and
> salvaged stone. . . .
>
> And, of course, storms like the ones we've been expe-
> riencing this year are predicted to become more common
> place, sea levels are rising, and we're living in very chang-
> ing times. The National Trust is still committed to the
> results of the 2005 Mullion Harbour Study, which states
> that we will continue to maintain and repair the harbour
> until a catastrophic event, at which time we embark on a

programme of "managed retreat." Fingers crossed, despite these ferocious storms, that time may not be with us yet.[28]

I went out to meet Bob again on March 27. As I approached the harbor, I came upon the stacks of salvaged stone, arranged in rough pyramidal heaps among the hauled-up fishing boats and crab pots. The stones were remarkably regular—rough-cut granite blocks, all the size of a large loaf of bread or a small shoebox. On each, one face was stained and scuffed by exposure and wear. The faces that had been sealed away in the harbor walkway were strangely unblemished and bright where the light reflected off their fine-grained surfaces. There were many separate stacks, some more carefully arranged than others. A neatly squared-off tower surrounded the base of the iron capstan, rising to chest height in tier after tier. Another deposit lay unceremoniously heaped in a low pyramid near the quay, partly obscuring the "Caution" sign.

Bob waited out on the harbor wall, holding an umbrella. A rainstorm had just passed through; the leat built into the west wall sluiced stream runoff into the high tide basin, and a long scarf of water cascaded over the far cliff. We began to talk, mostly about the past, as if in wary agreement to avoid more sensitive topics. Bob pointed out the old soapstone quarries on the far cliff ("just line up with the edge of the breakwater"), the kink in the walkway where they had attached the western breakwater to the old structure, and the spot in the inner harbor where the storms had exposed the wooden base to the original A-frame crane used during harbor construction.

As we walked around to where the southern breakwater had been gouged out, the crumbled core now visible, Bob reported on the latest developments regarding a box of leaflets (titled "Mullion Cove: A Strategy for Coping with Climate

Change") that had appeared on the damaged harbor walls a few weeks earlier. He and others had assumed that the leaflet had been drafted in response to the damage caused by the recent storms, and he expressed some concerns about its representation of both climate science and local context. The line in the leaflet that seemed to rankle most was this one about the harbor study decision: "Receiving strong support from the community, this plan allows residents and visitors alike to enjoy the harbour for as long as possible, but recognises that at an unpredictable point in the near or distant future, the cove will once again look like it did in 1890."[29] I mentioned that the leaflet had been in circulation several years ago, when I had started to research the harbor, and that I had quoted it in my article. If there had been "strong community support" when the decision was made, Bob implied, then it had eroded over time, and no one he'd spoken to recalled ever seeing the leaflet before. He mentioned that when he retired to Mullion in 2007 (after a first visit in 1972 and thirty-five years of returning frequently for holidays), people didn't talk much about the future of the harbor.

There was a sharp irony to the recent turns of events, in that the leaflet had been produced by the National Trust in 2006 after the conclusion of the harbor study to communicate the decision and to keep it alive in people's memories. I picked up a copy on that first visit to the cove in 2007. An interval of several years passed, when then leaflet went out of circulation and wasn't stocked, simply because there were always other, more pressing things to think about. The mysterious reappearance of the leaflet (as if newly minted) in the wake of an unprecedented storm season had then been taken as evidence of a conspiracy to exclude the local people from deliberations about the future of the site. There was something of a storm-in-a-teacup character to the whole misunderstanding, but the underlying anxiety was genuine. Where Bob saw arrogance and indifference on the part of the National Trust, I saw distraction and overburden, each a side of a spinning coin.

We left the harbor and drove up to Bob's house so that he could show me some newly discovered photographs of the construction of the harbor. Over tea and lemon drizzle cake, we examined the details of the western breakwater construction—the foreman's odd hat, the rubble-strewn work site with a fin of bedrock rising up where the heart of the structure would eventually be. He pointed out the cranes and winches that they had used to move the rock around, and the way the underlying bedrock "reef" was excavated and redistributed to provide the fill for the harbor. Another image showed the dimensional granite coping stones and the setts stacked up, waiting, on the growing harbor wall. I made a comment about the bedrock that lay under the breakwater, speculating about whether its remnants would be exposed when the structure came down, and he chided me gently: "We don't want to think that way." Before I left, I admired his Victorian watercolor of Mullion Cove, painted in 1877, before the piers were built.

After that visit with Bob, I kept finding myself thinking about the piles of salvaged stones collected at the top of the slipway. Their presence—mute, inscrutable—seemed significant, but I wasn't sure I could articulate why. I imagined the embodied labor that had placed them there: people going down into the harbor at low tide, in small groups, finding the lost stones, picking them up one by one, and moving them up the beach, out of the reach of the highest tide, holding them safe until they were needed again—stones gathered as ballast against an uncertain future. I felt an odd lack when I realized that

I'd missed the opportunity to help, although, in the midst of the storms, I had offered. The salvaged stones seemed to bear mute witness to the events of the past winter and to the curious limbo state that the harbor now found itself in. In her novel *The Winter Vault,* Anne Michaels writes, "The moment one uses stone in a building, its meaning changes. All that geologic time becomes human time, is imprisoned. And when that stone falls to ruins, even then it is not released: its scale remains mortal."[30]

The harbor's granite was quarried from the earth 120 years ago, most likely from a quarry on the southern edge of the Carnmennellis pluton, perhaps from the quarry that lies outside the village I live in. The setts would have been split off with "plugs and feathers" from a thick slab, then pitched with finer tools to square them up at the edges. They traveled to Mullion Cove, where they were set into the face of the walkway on the new harbor walls, entering "human time"—carts full of newly caught fish trundling over them, crabbers heaping their crab pots on their coarse surfaces, holidaymakers strolling along them. The sea periodically, perennially, scooped some of them up and tossed them into the cove, where they were collected, stockpiled, and reset, though some escaped.

Now, in the most recent phase of ruination, the escaped stones had been hauled back into the human realm once again. The stacks at the top of the slipway resembled chance cairns, in that they contained within them a memory of the harbor (as a built form around which a community gathered itself) as well as a memory of their origins (the raw geology that had produced them). The cairn is a curious kind of monument, accreted rather than built, and occupying an indistinct space that straddles constructs of natural and cultural being. The Gaelic root (*carn/cairn*) refers equally to a rocky hill or a heap of stones assembled by human hands. The rocky outcrops of Cornish cairns are the homes of the giants; Scottish cairns often mark the site of a burial or an abandoned village; but everywhere they are way markers, guiding the traveler through unfamiliar terrain. They are also living monuments that expand and contract over time; they are, as Paul Basu observes, "continually refigured as they gather, coalesce and are dispersed again."[31]

Perhaps the act of gathering the stones and placing them at the top of the harbor could be understood as a way of keeping open the possibility of the structure's recovery while also acknowledging its mortality, and in doing so gradually coming to terms with the specter of loss. The stockpiled stones gave witness to, in Judith Butler's terms, "the vulnerable or precarious nature of embodied existence" and made tangible "forms of sociality and belonging" linked to the recognition of this shared vulnerability.[32] The act of salvage formed a loose collective of people linked through their care for the harbor; this collective formation existed apart from any individual differences of opinion about the structure's desired future. Meaning arose from the encounter with the materials and the unscripted, instinctive impulse to recover what had been lost. By doing the obvious and necessary thing, the people involved in reclaiming the stones created an open monument. They performed preservation as a "provisional and situated response that makes the objects it requires."[33] The salvage occurred at the point where narrative failed.

I went back to the harbor alone on April 1, set on finding my own sett. The recovered stockpiles seemed larger than they had the week before. The tide was still dropping, and I went down to look within the harbor walls first, where I found only the scoured surface of the beach, stripped of its sand by the storms. Some of the sand had been cast into the rough tunnel that runs under the southern breakwater, the navigation of which used to involve a treacherous walk through pits and pools. I walked over the newly smooth sand floor and emerged on the far side of the outer harbor walls. There I found a stone of roughly the right size and shape in a shallow pool at the base of the walls, though its edges were worn and it looked to have been in the sea for a long time. I thought to carry it back through the tunnel, but I was stopped short by the weight of it, worried I'd slip on the way. Its heft had me thinking again of each salvaged sett, handled repeatedly—lift, carry, set, lift, stack. Yes, I thought, you can spin stories about how it was and how it will be, but stories won't cause the swerve that would stop us from caring, stop us from going down to the edge of the sea and hauling up the lost stones, one after another, as large as loaves and as heavy as hope. What to do when the world begins to fall apart around you? Try to hold it together. Because not to do so would be to risk indifference.

> There are three conditions which often look alike
> Yet differ completely, flourish in the same hedgerow:
> Attachment to self and to things and to persons,
>   detachment
> From self and from things and from persons; and, growing
>   between them, indifference
> Which resembles the others as death resembles life,
> Being between two lives—unflowering, between
> The live and the dead nettle. This is the use of memory:
> For liberation—not less of love but expanding
> Of love beyond desire, and so liberation
> From the future as well as the past. . . .
> . . . History may be servitude,

History may be freedom. See, now they vanish,
The faces and places, with the self which, as it could, loved
   them,
To become renewed, transfigured, in another pattern.[34]

I came back through the passage and went down to the turn of the tide, across the rucked and ridged harbor floor. I spied my sett at the base of the inset steps, a rectilinear glint in the low water, the squared corner just visible. I had to step out to pull it from the sea, and a wave sloshed over my trousers. On examination, it was clear that the granite had been in the harbor for longer than a few weeks; its surface was stained with coralline algae and etched with relief calcareous squiggles left behind by serpulid worms. It had been cast off in some other season and was already on its way back over the border, into the world of mute matter. But I hauled it up the cove anyway, over uneven ridges and through saltwater puddles, trying not to slip. I left it with its brighter kin in a salvaged stack, returned to its mortal life.

Six months have passed since my April salvage expedition. After a round of consultations with structural engineers and insurance company loss adjustors, in spring 2014 the National Trust decided to undertake a full repair of the battered harbor walls. Work started on replacing the paving setts from the western breakwater in May and was completed in August, buoyed on midway by a visit from a minor royal on a tour of Cornwall's storm-scarred communities. The replacement of the setts was considered a like-for-like repair, and Listed Building Consent was not required.

   The damage to the southern breakwater was more extensive, and in June, the National Trust submitted an application to the county council requesting permission to repair the collapsed sections with a combination of concrete and salvaged stone.[35] The plan of proposed works was made available for comment on the council website. In July, an officer from English Heritage sent a letter detailing some concerns with the proposed work on the listed structure:

> We understand that some concrete has already been used for repairs; however, in the likely reoccurrence of storms, we would raise concern about the continued practise of complete concrete replacement and its cumulative impact on the breakwater, as there is a risk that it could be completely rebuilt in concrete losing much of the evidential, aesthetic and historic value that the fabric provides.[36]

The comment relies on an understanding of heritage that locates value in the (presumed) original fabric of a structure. When drafting this advice, the English Heritage officer would have consulted the List Entry Summary that details Mullion Harbor's historic features.[37] The summary, which was entered in 1984 (shortly after the replacement of the failed concrete slope with a historic facsimile of the 1890s design), notes the connection to the fishing industry and ends with the assertion, "The harbour has remained largely unaltered." Perhaps, in part, because of the omission of the history of repair and reconstruction from the list entry description, contemporary concrete repair was seen as a direct threat to the "evidential, aesthetic and historic" integrity of the harbor walls.

For English Heritage, the link between material preservation and memory was sacrosanct: there would be no contemplation of an approach that performed memory through other acts of material engagement. Eliot asks us to contemplate the "use of memory" for liberation, "from the future as well as the past," and to accept that places will "become renewed, transfigured, in another pattern," carrying their history with them through those changes.[38] English Heritage frames the "use of memory" in a much more conservative sense. Despite the harbor's history of continual repair and rebuilding (and the doubtful nature of any claims to "original" stone), they insisted that the repair produce a simulacrum of stasis. In the compromise repair plans, agreed after consultation with the National Trust, concrete would be "coloured to match the surrounding stone with original stone," and repair with stone would be required on the visible walls of the inner harbor. To the extent possible, the harbor would be returned to its "timeless" appearance.

As I write this, the repair work is underway, but the newly imposed requirement to use serpentine stone on the inner walls has caused problems. All of the areas' serpentine quarries closed down years ago, and there is no source for new stone. The contractors have begun quarrying the heaps of boulders caught in the inner kink of the southern breakwater and around its seaward walls, looking to recover dimension blocks that were cast off in storms eight years ago—or eighty. The harbor will be reassembled, and the storms will come again, as they always have. Letting go is hard to do. Is that the moral of the story?

Do our stories need morals? I suppose where I am left at the end of this chapter is wanting something else—stories that open up rather than close down, stories that acknowledge that there are many ways of expressing attachment "to self, and to things and to persons," and that all of these are valid. In the messy world that we inhabit, Bob and I coexist as counterstorytellers, spinning around an axis that is Mullion Harbour. We don't want to prove each other wrong because that would sink the stories, so we reserve judgment and carry on. When we are no longer able to make sense, we are forced to acknowledge ambivalence, contradiction, care. The process of "loss adjustment" at Mullion has only just begun, and it will go on for decades. What is clear is that the decision made in 2006—"maintain and repair until failure"—was not a decision at all, but a deferral. Now the hard work begins. I need to leave Mullion now, and I will have to resist the urge to return to this place again. I can't still the changes that are going on there any more than I can predict them.

> Or say that the end precedes the beginning,
> And the end and the beginning were always there
> Before the beginning and after the end.
> And all is always now. Words strain,
> Crack and sometimes break, under the burden,
> Under the tension, slip, slide, perish,
> Decay with imprecision, will not stay in place,
> Will not stay still.[39]

Orderly Decay

PHILOSOPHIES OF NONINTERVENTION

> It was looking at something farther off
> than people could see, an important scene
> acted in stone for little selves
> at the flute end of consequences.
>
> *William Stafford, "At the Bomb Testing Site"*

A FEW HUNDRED MILES NORTH and east of Mullion Cove, on a bulge of Suffolk coastline that nudges into the North Sea, the National Trust is presented with another riddle in stone and concrete. Orford Ness (also known as Orfordness, the Ness, or the Island) is a long, crooked finger of salt marshland and stony shingle that extends south along the coast from a land bridge near Aldeburgh. This isolated landform has a peculiar place in British defense history. In the mid-1950s, as tensions with the Soviet Union tightened into the extended animosities of the Cold War, the United Kingdom's Ministry of Defence (MoD) pledged to develop an array of atomic weapons that would demonstrate Britain's commitment to strategic deterrence.[1] The MoD decided to establish a weapons testing facility in a location that had been intermittently occupied as a classified research and test site since the early part of the twentieth century. Construction of the Atomic Weapons Research Establishment (AWRE) facility at Orford Ness began in 1955 on a former bombing range at the southern edge of the spit, away from the concentration of structures left behind by previous occupations.

Over the coming years, the MoD built six large test cells to house specialist laboratories. The massive concrete walls of the first labs were banked with local shingle and roofed with lightweight corrugated steel. Later labs were roofed with massive reinforced concrete platforms (held up by blocky columns and heaped with more shingle) that were designed to implode and contain debris in the event of an accidental explosion.[2] The work carried out in the labs focused on testing weapons trigger mechanisms for safety and reliability. Inside the massive structures, bombs (absent their fissile nuclear core) were exposed to simulations of the environmental stresses they might plausibly encounter before detonation: vibration, temperature extremes, sudden shocks. Technicians controlled the tests remotely and used radiotelemetry to monitor the response of the devices. At the height of operations, the AWRE facility employed 350 personnel, though only a handful of these people were aware of the full scope of the work being carried out. Most employees, sworn by the Official Secrets Act, restricted their attention to the specific task they had been set—processing data, designing circuits, making gauges—unaware of the wider context within which they operated.[3]

AWRE continued its work on Orford Ness until 1971, when it shifted the test operations to its main base at Aldermaston, Berkshire. In the same year, sections of Orford Ness were designated as a National Nature Reserve. Despite decades of degradation brought about by the activities of the MoD, the Ness retained swaths of rare vegetated shingle habitat; the reserve designation was intended to protect these areas and encourage their recovery. The conversion of the landscape from military reserve to nature reserve was gradual and partial, however, and bomb disposal teams made frequent visits to defuse warheads and excavate live ordnance.[4] The massive lab structures remained in place, monuments to destruction deferred. The silhouettes of the Brutalist "pagodas," visible from the shore at Orford Quay, became hazy icons of Cold War secrecy; until the 1990s, the work carried out on the Ness remained strictly classified. When writer W. G. Sebald walked the Ness two decades after AWRE's desertion, he mused darkly, "I imagined myself amidst the remains of our own civilization after its extinction in some future catastrophe."[5] The site, which had avoided annihilation during its active use, was subject instead to a slow implosion brought on by vandalism and salvage, and gradual incursions by owls and gulls, roots and rust.

Orford Ness embarked on a new interval in its history in 1993, when the National Trust acquired the site from the MoD. The purchase was a departure for the organization, whose significant property holdings tended toward stately homes, open countryside, and undeveloped coastline, not degraded ex-military sites. Although the purchase of the site was heavily debated within the organization, eventually the conservation value of the marsh and shingle habitats swayed the decision. The relics of the military research facility on the site posed problems, however. Many of those involved recommended "tidying up" the landscape by removing the offending structures and allowing the place to be "converted back to a wilderness."[6] A few farsighted individuals, however, advocated for a more sensitive treatment of the site's derelict structures.

Angus Wainwright, the National Trust archaeologist responsible for the region, carried out a survey that described the aesthetic characteristics of three distinct "character areas" on the Ness, including the old airfield where most of the remains from World War I and World War II were concentrated, the former grazing marshes and the open shingle occupied by the AWRE structures. "One of the key aesthetic qualities" of the shingle area, observed Wainwright, was the "process of colonization and decay of the man-made by nature."[7] He described the powerful effect created by the "ruinous condition" of the former labs, and the sense of disorientation and disturbance generated by the monumental scale of the abandoned structures in the exposed landscape. The Statement of Significance prepared for the property at the time attempted to articulate some of the tensions that defined the site: "The site is characterised by contrasts: the man-made versus the natural, hard forms versus soft forms, past activity compared to the present stillness, and most significantly, the timeless natural process contrasting with transitory man-made dereliction."[8] After the survey, the National Trust decided to adopt a "general philosophy of non-intervention" for the site and its structures.[9] A subsequent management plan proposed a selective and targeted application of this approach. Many buildings from the site's early twentieth-century incarnation as an experimental test facility for aerial warfare and bomb ballistics research were near collapse: ultimately, twenty-seven structures—including a World War I–era station headquarters with parquet floors—were demolished. The National Trust rehabilitated and repurposed other buildings for their operations, and a handful of buildings, including an experimental radio navigation facility known as the Black Beacon, were marked for restoration.

The philosophy of nonintervention—also referred to as "continued ruination" by current managers—would be applied most comprehensively to the management of the residual AWRE structures. After an initial clearance effort, intended to remove toxic substances and major safety hazards, the Trust would carry out no regular maintenance, and nothing would be done to arrest (or exacerbate) their decline. In this way, decision makers hoped, the symbolic value of the structures would be retained, expressing "the role of technology in late twentieth century warfare and the awesome destructive forces it unleashed."[10] In Charles Merewether's terms, the structures would be allowed to function as "negative monuments . . . at the threshold between the impossibility of remembering and the necessity of forgetting."[11]

I visited Orford Ness twice in 2012, almost two decades after the National Trust's acquisition of the property. I was curious to learn how the philosophy of nonintervention had played out in practice and to discuss its implementation with the people who managed the site day to day. My first impressions were not of the ruined structures, however, but of the strangeness of the landscape itself. The part of the Ness owned by the National Trust is accessible only by boat, across a tidal channel. A ferry deposits visitors at the inland, marshy edge of the spit, and from there the AWRE lab structures, when they are visible at all, seem implausibly distant—faint, blocky features on the horizon. To reach them, you have to travel along the old military roads and across a series of low bridges; you are often forced to move at a tangent to your destination, heading away in order to cut back. The shingle ridges become more prominent as you reach the seaward side of the spit, their low spines etched in shallow, concentric arcs across the open expanse.

As you near the area once occupied by the AWRE test facility, the lab structures come into focus, resolving into discrete masses of concrete and shingle. Visitor access to this part of the site is strictly controlled, and people are advised to stay on the marked route. On approach, I passed a sign reading, "Warning: do not touch suspicious objects," illustrated with a stylized graphic of an exploding bomb, but the only suspicious objects I could see were spiky teasel plants. In Lab 1, the first building on the marked path, I entered a damp concrete passage that ended abruptly at a steel mesh barrier. Beyond lay a large open space, where high, moss-stained walls rose to a punctured ceiling of riblike girders. Flayed sections of roofing and ductwork hung above the flooded floor; the dark surface of one roughly rectangular pool along the right wall suggested some depth below. Vibration tests were once carried out in this pit. Around the pit's edges, tiny vegetated islands crowded with opportunistic dock, willowherb, and chickweed. The plants had taken root in and around material shed by the decaying structure: twisted roof panels, a bent lighting fixture, strewn sections of plaster and pipe. A door opening through to a second chamber was visible at the far end of the ruined space.

I left Lab 1 and walked along the cracked roadways across the exposed shingle. A scatter of rusty debris radiated out from the lab structures—snarls of oxidized wire and cable, sections of tarmac colonized by lichen, crumbling concrete anchors. Around the far side of one of the looming pagodas, Lab 5, I followed a dark flight of feather-strewn stairs down into a sunken chamber. The shingle-heaped, reinforced concrete roof was held up by rows of squat columns, and light entered through the high spaces between them to dimly illuminate a rectangular space. The floor was covered with loose shingle that had been tossed over the clerestory gap during seasons of storms. Rows of vertical steel plates were set in three high walls, each plate incised with slots that formed the shape of an elongated cross. These slots were once used for mounting equipment during vibration tests: "vibrated objects might also be placed in jackets to simulate extremes of heat and cold, or in a portable altitude chamber to mimic the effects of altitudinal changes."[12] In the eaves of this strange atomic chapel, doves nested on a precariously suspended section of metal ductwork.

The booklet I carried with me on my self-guided tour stated, "We aim in our management to preserve evidence of past use at the site and at the same time allow natural processes to run their course."[13] Preservation of the material past and accommodation of natural process are usually presumed to be incompatible aims, but here the two seemed to hold an uneasy truce. In an interview in 2009, site ranger Duncan Kent commented, "Our policy has been one of nondisturbance, allowing nature to reclaim the place, rather than to embark on a debris clearance programme. The sense of dereliction adds to the atmosphere."[14] The National Trust's philosophy of nonintervention couples an explicit appreciation of the aesthetics of decay with acknowledgment of the imperatives of ecology. While this fusion of aesthetic and ecological concerns is almost unheard of in mainstream Euro-American conservation contexts, there are echoes here of earlier deliberations about preservation practice, which circulated in the decades before legislation codified and constrained approaches to the material past.

In his 1903 essay "The Modern Cult of Monuments," Viennese art historian Alois Riegl identified three distinct "memory values" associated with acts of historic preservation: age value, historical value, and commemorative value.[15] Riegl was interested in how these different values came to be associated with cultural artifacts (which Riegl referred to as monuments) and how the assignment of certain values in turn presumed the application of particular modes of care. Commemorative value he associated with "deliberate monuments," which had been created with a memorial intention: "the fundamental requirement of deliberate monuments is restoration."[16] Historical value, in contrast, was based on the value of the monument as a representation of a stage "in the development of human creation," and as such its perpetuation required preservation of the artifact in its present state.[17] Age value he framed as entirely distinct from these two approaches.

When age value is given precedence, Riegl explained, cultural artifacts are treated as "natural organisms," and the "reign of nature, including those destructive and disintegrative elements considered part of the constant renewal of life, is granted equal standing with the creative rule of man."[18] In Riegl's formulation, the recognition of age value deferred the impulse to preserve or restore and resisted "unauthorised interference with the reign of natural law."[19]

> If the aesthetic effect of a monument, from the standpoint of age value, arises from signs of decay and the disintegration of the work's completeness through the mechanical and chemical forces of nature, the result would be that the cult of age value would not only find no interest in the preservation of the monument in its unaltered state, but it would even find such restoration contrary to its interests. The modern viewer of old monuments receives aesthetic satisfaction not from the stasis of preservation but from the continuous and unceasing cycle of change in nature.[20]

The valuing of dissolution and material change operates through what David Lowenthal has described as an "aesthetics of rupture," in which fragmentation and disintegration open up a connection to the past by attesting to the passage of time and shocking "the viewer into a double apprehension, of its presumed original state and of its ineluctable decay." [21]

At Orford Ness, what Riegl described as "the unceasing cycle of change in nature" is everywhere evident. Inside the lab structures, "mechanical and chemical forces" have free play: interior walls weep and spall with the symptoms of concrete "cancer," a condition traced back to the salt water used in the original mix. Subsequent infusions of salt spray have rendered metal components equally vulnerable: bolts and beams swell with rust, then erode in brittle flakes. Processes of colonization by plants and animals produce novel ecosystems in unlikely locations. A tiny English field has established on the top of the crumbling brick wall that once contained the centrifuge facility in Lab 2. Hawkweed, false oat, and red fescue march along the narrow sill in linear formation and work their roots into the crumbled pointing.

While Riegl's classifications of conservation value are often invoked by contemporary scholars, few have specifically addressed the radical processual orientation that underpins his theory of age value. Cornelius Holtorf draws on Riegl's ideas to propose that the evocation of age value relies on the perception of what he calls "pastness," an effect enhanced by visual evidence of erosion, decay, and deterioration. Holtorf proposes, however, that pastness is never inherent in an object and connected to its material substance but is instead the result of a given perception of an object in a given context, even if the object happens to be a reconstruction or a simulation.[22] The physical substance of the object, in this formulation, is secondary to the social and cultural context in which it is located. Stephen Cairns and Jane M. Jacobs engage more directly with the challenge that the theory of age value poses to conservation practice, observing that Riegl's explicit framing of decay as a productive architectural contribution achieves a reassignment of the "authority of creativity" from the human hand to the "lawful" forces of nature.[23]

His fascination with "nature's law of the transition of growth into decay," as well as his insistence on witnessing the "life cycle of the monument," suggest a commitment to process that goes beyond mere patina:

> Modern man at the beginning of the twentieth century particularly enjoys the perception of the purely natural cycle of growth and decay. Thus every work of man is perceived as a natural organism in whose development man may not interfere; the organism should live out its life freely, and man may, at most, prevent its premature demise.[24]

Writing in 1903, Riegl perceived an openness to the accommodation of entropy that, within a decade, had begun to close down (although the actual extent of the aesthetic appetite for decay seems, perhaps, to be overgeneralized in his sweeping reference to "modern man"). The passage of heritage protection legislation and the increasing professionalization of conservation as a field of practice began to calcify the new conservation paradigm, with its emphasis on the preservation of material fabric.[25] Certainly the potential for explicit acknowledgment of ecological process and material change as a positive element in heritage practice, rather than an unfortunate inevitability, has not been fully realized since. But Orford Ness provides a rare glimpse of how it could have been otherwise.

The philosophy of nonintervention at Orford Ness has resulted in some interesting challenges for the people who manage the site: they seek to "allow natural processes to run their course," but in doing so, they must adapt their interpretation to a continually shifting mosaic, in which things transgress and trouble the classifications usually used to ascribe significance to "natural" and "cultural" heritage. For example, many of the plant species that have rooted in the ruined lab structures are described as *ruderal*, a term used by ecologists to describe plants that grow on disturbed ground. The term derives from the Latin *rudus*, "rubble" or "broken stone." Ruderal species are often found on road verges and building sites, wastelands and waysides. The seeds of the dock and chickweed that have colonized the damp, sheltered corners of the lab structures may have blown in on an offshore wind or hitched a ride on the shoes of the bomb disposal team. But the distinction between the weed species that have colonized the ruined lab structures and the native plants that live on the shingle is not so clear-cut. The yellow-horned poppy (*Glucium flavum*) is a pioneer species that establishes itself in newly accreted shingle ridges. Loss of shingle beach habitat (due in part to reinforcement of coastal defenses) has led to a decline in U.K. populations over the last few decades, and disturbance of the wild plants is prohibited. Some ecologists would describe the yellow-horned poppy as a ruderal species, in that it is adapted to thrive in a mobile and dynamic system, continually reshaped by wind and weather. On Orford Ness, these poppies have colonized the shingle ridges, but also the ruined lab structures. The wild, native species thrives alongside mundane and often maligned weed species by dint of their shared preference for disturbed, unstable substrates.

For the past twenty years, conservation work on Orford Ness has been carried out under the auspices of the European Union nature conservation program. The overarching aim has been to "reconcile past military experimental use with the requirements of nature conservation."[26] Activities have involved experimental restoration of an area of degraded shingle, winter flooding of the grazing marshes, and creation of brackish water lagoons to improve habitats for breeding, overwintering, and migratory bird species. As conservation goals have gradually been met, protective designations have thickened: Orford Ness is now recognized as a National Nature Reserve, a Site of Special Scientific Interest, a Special Area of Conservation, and a Special Protection Area. Within this welter of designations signaling exceptional nature conservation value, the status of the unexceptional species that have colonized the derelict fringes of the AWRE site is unclear. They are tolerated, but there is little attempt to narrate their story as part of the gradual transformation of the site by "natural processes."

Opportunities exist, however, to tell stories that stitch together the cultural history of the site with its evolving natural history. Areas of the former AWRE site support vigorous stands of rosebay willowherb, its distinctive purple spears rising up to puncture the silhouettes of the brooding lab structures. Rosebay willowherb (*Chamerion angustifolium*) is also known by the names fireweed, bombweed, and (strikingly) ranting widow, because it is often the first plant to take hold in the wake of devastation by fire or war. Richard Mabey writes of how "the summer after the German bombing raids of 1940 the ruins of London's homes and shops were covered with sheets of rose-bay, stretching, according to some popular reports, as far as the eye could see."[27] In the aftermath of the bombing of Hiroshima, the counterpart pioneer plant was the sickle senna, a tall, leguminous herb with yellow flowers. A survivor of the bombing, on leaving the hospital after a month's stay, described what she saw to writer John Hersey:

> Over everything—up through the wreckage of the city, in gutters, along the riverbanks, tangled among tiles and tin roofing, climbing on charred tree trunks—was a blanket of fresh, vivid, lush and optimistic green; the verdancy

rose even from the foundations of the ruined houses. . . . .
Especially in a circle at the center, sickle senna grew in extra-
ordinary regeneration. . . . It actually seemed as if a load of
sickle-senna seed had been dropped along with the bomb.[28]

At Orford Ness, a stand of rosebay willowherb, juxtaposed
against the dark bulk of the lab structures, where the specter of nuclear annihilation was seeded and cultivated for two
decades, functions as a living, concrete poem. With a little
effort, we can see the ruined cities that existed only in our
imaginations, the shadow of what could have been.

The ruined state of the lab structures can also, however,
lull us into thinking that this particular chapter in history is
over, when in reality the centers of nuclear calculation and
proliferation have only shifted or gone underground. To return to the Statement of Significance for the site:

The buildings at Orford Ness can be looked at as part of the
documentation of past events, as symbolic of deep-seated
urges within our culture, or merely as dramatic forms in
the landscape. . . . They also say a lot about our confrontation with the forces of nature and the ability of these forces
to adapt our structures and given time, destroy them.[29]

In an interesting reassignment of agency (and an echo of
Riegl), in this formulation the "forces of nature" take on the
burden of destruction, deflecting attention (perhaps) from
the other, more sinister, forces of destruction implicated in
this place and naturalizing a brutal legacy. To the extent that
a particular historical moment is allowed to recede through
the gradual dissolution and decay of its material imprint, one
could argue that nature is being used to obscure or excuse the
site's uncomfortable past.

Or perhaps the juxtaposition of the forces of nature with the implied atomic force contained within the structures does something else, allowing us to acknowledge our ambivalence about places like Orford Ness and stay the urge for resolution and explanation. The management of other postmilitary landscapes is often characterized by an impulse to tidy away the legacy of violence and destruction, to "militarise the natural and naturalise the military."[30] Jeffery Sasha Davis has written about the "double erasure" that defines the reinscription of many militarized landscapes: "First there is an erasure of the social life that existed in the place prior to its takeover by the military. Second there is an erasure of the history of the military's use."[31] Unlike other postmilitary landscapes, where "weapons to wildlife" conversion[32] often removes physical evidence of military occupation, at Orford Ness, a deliberate decision was made to accommodate the contrasting presence of the ostensibly wild and the aggressively technological, the benign and the destructive.

The accommodation of these uneasily paired presences in the landscape can be perplexing for visitors, and the National Trust has not rushed to dispel the sense of confusion and contradiction that the site can generate. On disembarking from the ferry, every visitor is told, "A visit to Orford Ness should be safe, but not necessarily comfortable." The cultivation of dis- and misinformation is also part of the interpretive strategy at the site, and the resulting air of incongruity, coupled with the spectacular forms of the ruined structures in the landscape, has an unusual aesthetic appeal.[33] Early in their tenure, National Trust managers recognized that the qualities of the site presented a unique opportunity to integrate contemporary art practice into their activities. Painters John Wonnacott and Dennis Creffield were both invited to spend time on the Ness in the early 1990s. According to

Jeremy Musson, the images they produced of the eerie, abandoned structures in the exposed landscape informed the
management approaches that were subsequently adopted.[34]
Over the past two decades, dozens of photographers, painters, performers, and filmmakers have traveled over on the
passenger ferry to make work in response to the place's en-
crypted and enigmatic charms.[35]

In 2005, Louise K. Wilson spent several weeks at Orford
Ness creating a series of audio and video works that responded
to the atmosphere of "secrecy and strangeness" and to the
sounds associated with the renewed "occupation" of the site—
pigeons trapped in an air duct, a hare skittering over a shingle
bank, the creak of dislodged iron sheeting.[36] She placed her
work within the derelict structures to draw attention to the
acoustics of the abandoned spaces and to play on the site's historical association with processes of "transmission and reflection." In the summer of 2012, the National Trust commissioned
new work as part of a project called *Untrue Island.* Writer Robert McFarlane wrote a libretto, set to music by jazz musician
and composer Arnie Somogyi, and sisters Jane and Louise Wilson created a series of installations and sound works in the
AWRE structures. The Wilson sisters' *Blind Landing* was part of
a series inspired by the yardstick measures used to determine
scale in film sets.[37] In 2014, Anya Gallaccio created an installation that began with a single pebble scarred by a controlled
explosion, which she then crushed further before photograph-
ing its magnified remains. In her discussion of the piece, she
references her desire to express the "essence of the place and its
traumatic history."[38] As with other ex-military sites, the site is
attractive to artists who seek to, in Matthew Flintham's terms,
unconceal the "dark immanence of destructive potential in the
British landscape" and make present that which had been assumed (or hoped) absent.[39]

But the work of these artists also makes a clear contribution to the wider management philosophy of nonintervention. By responding to (and reflecting back to its audiences) the abandonment and dereliction of the site, site-specific art at Orford Ness works indirectly to validate benign neglect as a legitimate management strategy. Much of the work produced explicitly acknowledges and exploits the aesthetic attraction of decay—or, as Angus Wainwright describes it, the "order in disorder and the beauty in ugliness."[40] By making an asset of incompletion and fragmentation, such work allows the managers of the site to demonstrate the perceived cultural and artistic value produced by their hands-off approach. Some of the work (such as Gallaccio's) draws inspiration directly from the instability and latent volatility of the site, working from an "informed appreciation of processes of dissolution innate to both organic and inorganic nature."[41] Other work uses the spectacular ruination of Orford Ness site as foil or backdrop, and does not engage directly with the specificity of the site and its formative processes.[42]

Most of the artistic engagements with Orford Ness are transient and leave little trace in the landscape. As temporary interventions, however, they call attention to the radical potential for continued ruination as a heritage management practice that generates both ecological and cultural benefits. The work that is happening in this site can be understood, perhaps, as an expanded form of reconciliation ecology. In scientific circles, reconciliation ecology is defined as the branch of ecology that studies ways to encourage biodiversity in human-dominated ecosystems.[43] I wonder if the term could also be stretched, however, to acts of cultural and social reconciliation, in which we are faced with the dark and destructive elements of our past, and forced to acknowledge their continued potency without forcing their premature resolution. At a place like Orford Ness, a collaborative and compromised management ethic has gradually emerged; managers accept the inherent messiness of the landscape and do not attempt to create pure zones for either nature conservation or historic preservation. But this has required a willingness on their part to accommodate some measure of both semiotic and ecological autonomy—and to stay the attempt to control processes of both meaning making and material change.

Geographer J. B. Jackson, in his essay "The Necessity for Ruins," writes about the "interval of neglect" that must elapse before certain elements of the past are elevated to the status of heritage and protected as such.[44] At Orford Ness, there is some evidence that this interval may be coming to an end. In 2003, Wayne Cocroft and Roger Thomas published the results of a major study on the United Kingdom's Cold War heritage, and the facilities at Orford Ness were noted for their significant contribution to national defense technology.[45] The study coincided with a wider intensification of public and scholarly interest in the legacy of the Cold War, with amateur bunkerologists and academic archaeologists collaborating in the revaluation and reappraisal of these formerly marginalized sites (though this very marginalization had fostered a sustained appreciation during the intervening years among those inclined to seek out such places).[46] English Heritage published a comprehensive historic and archaeological survey of the AWRE site in 2009,[47] and in 2014 the test buildings and associated structures were scheduled under the Ancient Monuments and Archaeological Areas Act (1979) as a "monument of national importance." The potential for UNESCO World Heritage Site designation has also been mooted.

Now that the AWRE structures are formally designated

and subject to statutory protections, deliberations about
the site's philosophy of nonintervention will be back on the
agenda. There has been some discussion about the prospect
of a Heritage Partnership Agreement that would provide the
flexibility needed for the National Trust to continue to pursue
their unorthodox management practices. It is not clear, however, how such an agreement would be interpreted, and the
use of heritage protection legislation to authorize the continued degradation of a heritage resource may introduce a set of
contradictions too profound to remain unchallenged.[48] The
scheduling brings the AWRE structures under the oversight
of English Heritage (now Historic England) for the first time,
and it is likely that pressure will build on the National Trust
to stabilize selected structures and—belatedly—arrest their
decay.[49] In Riegl's terms, Orford Ness is at a critical juncture
where a decision must be made about whether to privilege
historical value over age value.

"Whereas age value is based solely on decay, historical
value seeks to stop the progression of future decay, even
though its entire existence rests on the decay that has occurred to the present day," Riegl wrote.[50] The AWRE structures in their current state provide a vivid index to the time
that has elapsed since their abandonment, a record evidenced
in rust and rot, and in the incursions of new growth among
the ruins. If the structures were to be stabilized, some attempt might be made to preserve the aesthetic of decay and
dereliction, but ongoing process would need to be arrested
in the interests of long-term preservation. Recognition of
historical value aims "for the best possible preservation of a
monument in its present state; this requires man to restrain
the course of natural development and, to the extent that he
is able, to bring the normal progress of disintegration to a

halt."[51] This is a prospect that the Orford Ness managers have contemplated. In the assessment of Grant Lohoar, a National Trust employee who has worked at the site since its acquisition, the AWRE buildings are still "retrievable." He accepts that there could come a moment when active preservation efforts are initiated, commenting, "We're not interfering with that possibility."[52] Others in the organization are more outspoken about the value of the current management philosophy, and point out that the site's historic and archaeological detail is preserved in extensive surveys. Angus Wainwright insists that the "ongoing process of the structures' decay, which in the view of the National Trust is such an important part of the aesthetic interest of Orford Ness, should be allowed to run its course."[53]

In a peculiar paradox, Orford Ness is a site threatened not by destruction but by preservation—the ruination of the ruin. "Objects framed as ruins need our attention and care because they are always threatened by loss, but if we care for them too much their status as ruins is threatened," observes Michael Roth.[54] The impossibility of reconciling the necessity for structural treatment with retention of authentic evidence of time's passage is, of course, one of the formative tensions embedded in modern historic preservation practice.[55] John Ruskin was perhaps the first to articulate this tension and to advocate for the virtues of selected neglect.[56] Riegl extended this argument in his 1903 essay:

> From the standpoint of age value, one thing is to be
> avoided at all costs: arbitrary human interference with the
> state in which the monument has developed. . . . The pure,
> redeeming impression of natural, orderly decay may not be
> diminished by the admixture of arbitrary additions. . . . The

> cult of age value . . . condemns every effort at conservation,
> every restoration, as nothing less than an unauthorised
> interference with the reign of natural law.[57]

What distinguished Riegl's ideas from Ruskin's was his insistence on examining and calling attention to the ecological and chemical processes that produced the cherished aesthetic. His conception of age value moved beyond appreciation of surface patina to follow processes of disintegration as they rearranged the monument's substance, and he found value in these processes in their own right. The past twenty years at Orford Ness have given us a sense of what application of Riegl's theory of age value might look like in practice, and if it is allowed to play out, the opportunities for further interpretation of the gradual intermingling of cultural and natural material will continue to multiply until "the unhampered activity of the forces of nature will ultimately lead to the monument's complete destruction." Riegl proposed that, over time, as decay progresses, the "age value of ruins becomes less extensive . . . [and] becomes more and more intensive, since the remaining elements have a more forceful effect on the viewer."[58] Intensive value centers on the force of the fragment, the remnant, the incongruous juxtaposition of that which persists against that which has been absorbed into other orders.

Even Riegl acknowledged, however, that our willingness to allow destructive process to run unchecked may ultimately come up against our reluctance to countenance complete loss: "This process also has its limits, for if the extensive effect of age value is lost completely, no substance remains for intensive effect. A bare, shapeless pile of stones will not provide the viewer with a sense of age value."[59] So we are left back in the midst of the old riddle. Riegl seemed most willing to accept limited and strategic intervention when "the forces of nature . . . threaten an abnormally rapid disintegration of its organism."[60] He questioned whether adherence to the "reign of natural law" is always strictly necessary:

> To the proponent of the cult of age value, a gentle interven-
> tion by the hand of man seems the lesser of two evils when
> compared with the violence of nature. In such cases the
> interests of both values would seem, at least on the surface,
> to go hand in hand, even though age value seeks merely to
> slow down disintegration, whereas historical value opts for
> a complete halt to the processes of decay altogether. The
> main issue for contemporary monument preservation is to
> avoid a conflict with both values.[61]

A century and a bit later, little seems to have changed. It is tempting to imagine that we may finally be ready to push past this paradox and to allow one place—already well on its way—to go on changing, and learn how to make sense of it in its going.

5

A Positive Passivity

ENTROPIC GARDENS

Petrol and diesel will both dry up
But that doesn't happen to a buttercup.
Flowers shoot upwards with mighty heaves
And sprout in a flurry of stems and leaves.
Here they come shouldering through the road,
Willowherb, Woodruff, Woundwort, Woad.

U. A. Fanthorpe, "Under the Motorway"

THE NARROW-LEAVED RAGWORT (*Senecio inaequidens*) has a long, spindly stalk and, as its name suggests, narrow green leaves. I encountered the plant on a visit to a former ironworks in the far west of Germany, where it grows abundantly and indiscriminately, on abandoned railroad grades and in the seams of spalling concrete walls. When I visited in July 2014, its slightly scraggly yellow flowers were in full bloom. In ecological terms, the narrow-leafed ragwort is a neophyte, a newcomer. This particular species of ragwort is native (and here we encounter the embedded politics of ecological etymology) to the far south of Africa. Ragwort seeds are thought to have arrived in the area in shipments of iron ore, which were transported up the Rhine to feed the massive blast furnaces whose skeletal structures the plant has now colonized. The ragwort is one of many traveling species that have established themselves in this postindustrial landscape, finding an unlikely refuge in the scarred and contaminated soils left behind by eight decades of intensive production and processing. Here it is part of a unique recombinant ecosystem that has developed in the three decades since production ceased and other processes emerged to alter and occupy the space.[1] That it has been allowed to thrive here, and to occupy the derelict structures, is testament to the unconventional management practices that have been adopted in this place.

In the last chapter, I visited a place where a policy of continued ruination has guided management for two decades and where intervention has been, for the most part, minimal— although there is some evidence that this interval in the history of the place may be coming to an end. In this chapter, I explore two places that have proposed or staged experiments in partial, rather than total, ruination. In both of these places, there has been some attempt to balance an accommodation of gradual structural decay and spontaneous renaturalization with a level of intervention required to sustain selective reuse and managed public access. At one site, this experiment has been running for twenty years—long enough to allow for some reflections on whether its original intentions have been realized. At the other, intentions are still conceptual, captured in documents and declarations of intent, though the shape of future orientation is discernible in early actions to manage the site. These places are linked through their adoption of shared principles of openness, iteration, and incompletion. They are both experimental sites—not in the sense that they have constructed a space where they control specific processes in order to test hypothetical outcomes, but in an alternative sense, recently articulated by Jamie Lorimer and Clemens Driessen, that understands experiment as a "tentative procedure; a method, system of things, or course of action, adopted in uncertainty."[2] In this sense, these are sites open to the workings of entropic possibility. But as exploration of these places shows, the embrace of entropy often slips in the translation from principle to practice.

Duisburg Nord Landschaftpark occupies a 570-acre site in Germany's Ruhr region in the far west of the country. The Ruhrgebiet, as it is known, is an area associated with intensive twentieth-century industrial production and, more recently, bold initiatives to recuperate and reinvent the extensively altered landscape as *Industriekultur* (usually translated as "industrial heritage").[3] Iron production at Duisburg began in 1903 and ceased in 1985. After closure, the site was the subject of an international design competition (as part of the Emscher Park International Building Exhibition) and reopened to the public in 1994 as a landscape park, designed by architects Latz + Partner. In this most recent incarnation, Duisburg has been the subject of sustained international interest, primarily focused on the way the site's design incorporates an unprecedented openness to reuse and regeneration (both social and ecological). The other site under consideration in this chapter is a derelict Victorian estate (Kilmahew) surrounding an iconic modernist seminary (St. Peter's) located west of Glasgow, Scotland. The seminary also was abandoned in the mid-1980s, but its subsequent recycling and repurposing has been slower to take hold. Several years ago, the public arts organization NVA (an acronym for *nacionale vitae activa*, a Latin term meaning "the right to influence public affairs") began to explore options for the site's future, and in 2010 they were invited to host a debate at the Twelfth Venice Architecture Biennale. They invited Tilman Latz, a current partner in the firm that designed Duisburg Nord, to take part in the conversation. By aligning themselves with the innovative experimental practices that had been trialed at Duisburg, NVA made clear its intention to adopt strategies for the Scottish site that also privileged indeterminacy, reinvention, and iterative participation. In a statement of intent published after the Biennale, NVA asserted that their plan for Kilmahew/St. Peter's would "embrace an incremental process of change" that "supports the principle of an *unfinished* work and accepts a level of entropy."[4]

What does it mean, in practice, to "accept a level of entropy"? In such a statement, entropy is invoked in the abstract, as a force capable of catalyzing an indiscriminate unmaking of material substance. In order to perform, however, entropy must work through a series of agents, and in a built structure, these agents come in many animal, vegetable, and chemical forms. Wildlife, weeds, and water initiate the transformative "micro-processes of wear and tear" that ultimately produce an entropic effect: oxidation, biodeterioration, percolation, cryoperturbation, erosion.[5] In a counterpreservation heritage practice, any attempt to find a balance between continued process and planned use must negotiate with these agents. As will become clear in this chapter, how people carry out this negotiation—and their willingness to accept the inherent unpredictability that entropy engenders—exposes a tension between intended conceptual innovation and its realization in practical applications.

Studies of ruins and ruination (both historic and contemporary) have been concerned in various ways to call attention to the way plants generate particular aesthetic effects as they take hold in disused structures, softening the hard lines of architecture and contributing to its "pleasing decay."[6] In such writings, plants are most often taken to be part of a generic collective of greenery, bundled into an aggregate identity. A similar lack of specificity also emerges in applied contexts when, for example, landscape architects or ecologists discuss the need for "vegetation management." The plants being managed (or, most often, removed) are rarely distinguished from one another, unless it is to demonize particular species as undesired invaders and thus further justify the need for eradication. Research on human–plant geographies has begun to explore this tension between collectivity and individuality, as well as the cultural reluctance to assign agency and sentience to individual plants.[7] This work has clear relevance in the historic environment context, where attitudes toward plants are characterized by ambivalence, in turn appreciative (of their aesthetic contribution) and apprehensive (about their role in accelerating deterioration).

The two sites explored in this chapter express subtly different attitudes toward the plant species that share their sites. At Duisburg Nord, plants appear to be treated as individuals, with their own complex ecocultural histories, and their own specific desires and demands. At the Scottish site, there is evidence of a more anxious and unresolved response. At base, as Michael Pollan explores in his book *Second Nature*, the story that people tell about the plants they share their space with is a story about control, about different levels of willingness to accept vegetative autonomy.[8] Attitudes toward herbal others can be placed on a continuum that runs from cultivation and containment to a looser form of collaboration and coexistence.

In articulating his original vision for Duisburg, Peter Latz identifies two central themes. First, he was concerned to allow the site to express its "physical nature" by accommodating ongoing "natural physical processes" such as erosion, oxidation, and spontaneous succession, and by creating new elements also subject to the working of such processes. The second theme he expresses as "utilisation," in that he sought to allow for the "metamorphosis of industrial structures without destroying them."[9] Many areas were to be "left to develop on their own without intensive treatment,"[10] including zones of persistent contamination, where soils and slag heaps were laced with deposits of heavy metals and hydrocarbons. In other areas (including a complex of former "sinter bunkers"), contaminated sediments were capped and sealed, and the new surface was planted with ornamental gardens. His approach sought not restoration (of the landscape to an imagined state of prior ecological equilibrium) but reconciliation, between the damaged landscape and unspecified future uses. Latz states, "The new vision should not be one of 're-cultivation,' for this approach negates the qualities they currently possess and destroys them for a second time. The vision for a new landscape should seek its justification exactly within the existing forms of demolition and exhaustion."[11] Rather than extinguishing the site's "aberrant processes,"[12] Latz sought to work with them, to allow the discourse between old industrial forms and emergent natural process to go about, in his words, "creating values between art and nature in a way that could never be made by the artist or nature alone."[13]

In this assertion, one can hear the echo of Georg Simmel, whose 1911 essay "The Ruin" identified a tension between form and process, or "nature and spirit," in the ruined structure. Architecture, he argued, is a sublimation of the inherent properties of matter—the durability of stone, the rigidity of metal—to serve the purpose of design and function. But in the ruin, when a building is subject to forces of "weathering, erosion, faulting . . . and the growth of vegetation," there is a moment when a precarious aesthetic balance is achieved: "Out of what art still lives in the ruin, and what of nature already lives in it, there has emerged a new whole, a characteristic unity."[14] Something like this "new whole," which emerges out of a relation between extinguished human function and resurgent natural process, is hinted at in Peter Latz's evocation of collaboration between the artist and nature at Duisburg (though Latz's approach arguably abandons Simmel's humanist hierarchical model to adopt a more radical symmetrical model of agency).[15] Russell Hitchings picks up on similar themes in his proposal that we shift from a focus on "material culture" to a study of "cultured materials," which examines the processes through which people impose their designs on physical matter and the ways in which that matter works back on these designs to express underlying "individual physical propensities" in unpredictable ways.[16]

I visited Duisburg Nord because I was curious to see for myself how Latz's original vision had been realized.[17] As at Orford Ness, I was drawn to the site's declared and deliberate accommodation of transformative process, an approach that seemed to unhitch the work of cultural memory from an insistence on the maintenance of durable objects. The background reading I'd done in preparation for my visit had seeded a certain skepticism about what I would find there. Most academic work seemed to hinge on analysis of the site as a study in polarized states: "intervention and neglect,"[18] "spontaneous and designed nature,"[19] "devastation and ecological reclamation."[20] Kerstin Barndt had written critically of the intentional dialectical tension that Duisburg Nord cultivates, observing, "The organic appeal of this dialectical move . . . smooths out the rough edges of the processes, and the overall sense is one of theatricality, in which visitors act out unscripted parts."[21] I was led to believe that I would find a place in which process was mostly patina, and where the promise of ongoing metamorphosis and dissolution was diluted through the prosaic imperatives imposed by public use of the site. In a sense, this is what I found—but I also found something else, which fits uneasily with this critical take.

I began my journey with a flight from Bristol to Amsterdam, where I spent the night in a budget hotel near Schiphol. On my way to breakfast, I noticed that the hotel escalator was stamped with the name of the company that had manufactured it: ThyssenKrupp. The recognition forced a spatial and temporal folding. I was already aware that the Duisburg Nord complex had been constructed by August Thyssen, one of the Ruhr's most prominent industrialists, to produce pig iron for processing in his steel plants. A Web search back in my hotel room revealed that in the 1990s, Thyssen AG merged with a company founded by Alfred Krupp, another regional steel magnate, to form ThyssenKrupp, a company that describes itself as "a diversified industrial group with traditional strengths in materials and a growing share of capital goods and services businesses."[22] The trip began with a reminder that postindustry is always site specific, and that industrial process is often displaced rather than suspended.

As we approached Duisburg station, buddleia and willow-herb flourished along the tracks, overgrown even on the active lines, completely obscuring the disused ones. I caught a tram to the north of the city and entered the park on its south side. A row of interpretive signs marched alongside the path into the main entrance, steel slabs over six feet high, printed with maps and text in both German and English. One of them offered this passage by way of introduction:

> The "Sacro Bosco" (sacred grove) in the medieval Italian
> town of Bomarzo is a renaissance park containing statues
> of fantastic and fabulous animals, surrealist figures, giants
> and monsters. Similar associations were the inspiration for
> the designer Peter Latz in his work on the North Duisburg

Landscape Park. The disused ironworks used to spit out huge gobbets of fire into the night sky, but the giant has now been tamed and lies exhausted. The site was formerly a no-go area for the general public, but its gates have long been thrown open to visitors. Rusty red blast furnaces and pitch-black storage towers, a labyrinth of pipelines and bunkers for coal, ore and other materials have been transformed into an exciting leisure area with attractive gardens and open viewing areas.

The North Duisburg Landscape Park is one of the key works in the Emscher Landscape Park because it symbolises a specific approach to developing disused industrial areas. This was the first venue where attempts were made to integrate industrial nature and industrial relics in a unified design and make their unique features and different chronological periods visible and comprehensible to the general public.

The areas of the park nearest to the entrance were almost completely dedicated to the repurposing of the site as an "exciting leisure area." Couples canoodled on benches in the open plaza between the hulks of the two blast furnaces; an area between the gasometer (now a diving facility) and the former casting house (a cinema and concert venue) was set up as beer garden with palm-frond kiosks, a white sand floor, and an inflatable Pilsner bottle. Around the corner I came across the *Piazza Metallica,* a grid of iron plates installed as part of the original park design and intended to express "erosion by natural physical processes" (though the space is now also used for temporary toilets and Dumpsters, which tempers its aesthetic unity somewhat).[23]

Slightly farther from the center, the uses of the site became more loose and improvisational: teenagers roamed in packs and left their traces (love locks, beer bottles) behind in odd corners and ostensibly off-limits areas. Climbers scaled the walls of the pitted concrete bunkers (one of the climbs entitled "Monte Thysso"), and artists daubed en plein air portraits of tangled steel and brooding machinery. The mass of one blast furnace was open to allow the public to wander up through its oxidized innards to a viewing platform, seventy meters above the ground, past various interpretive panels (which few people seemed to be reading). The view from the platform was instructive; the whole horizon to the west, along the Rhine Valley, was crowded with belching smokestacks and massed industrial buildings, obviously still active and productive. In this wider context, the park came into focus as anomalous and premature, promising to heal the scars left by industry with a soothing layer of greenery and an array of diverting leisure activities. From this elevated vantage point, the dominant impression of the park itself was of a place unevenly consumed by a creeping blanket of leaves and vines, which seemed to be encroaching from the outer edges inward, toward the relatively clear zone immediately surrounding the blast furnaces and the repurposed structures.

As I descended the blast furnace, I noted areas of rust and regrowth: patches of moss, tiny willow saplings rooted in pipes, flat roofs furred with green. At the base of the buildings, there was evidence of more active incursion—though primarily in inaccessible areas behind temporary fencing, where thickets of buddleia had sprung up and brambles were casting out straggling, speculative arms in search of new territory. Where the plants reached through the fences, they had often been pruned back. On one fence, a section of buddleia trunk as thick as my wrist had fused into the grid of steel, and although it had been cut off at either end the desiccated remnant held on, witness to an attempt to control and contain. Other obvious efforts had been made to check growth and re-generation. Rotting stumps resprouted with new shoots, and some patches of ground seemed to be artificially suspended in a stage of early successional regrowth. But slightly farther from the core, the situation was more ambiguous. Clumps of ragwort were firmly rooted in cracks in a concrete wall, above a well-traveled path. Ivy fingered its tough roots into ledges and loose sections of masonry, and saplings thrived along the tracks of an elevated railway. It was not entirely clear how much management was taking place, or even if anyone was paying much attention to the gradual erosion of structural integrity—and the attendant risks to public safety. There was not a total abdication of control here, but something more uncertain.

The ambiguity was enhanced by the contrast between the unruly edge lands and the designed garden spaces. In one bunker garden, serried rows of lavender held in tight formation, and another was carefully planted with roses. A third contained a clipped boxwood hedge trimmed into wavy stripes, interspersed with rows of hydrangea. These gardens seemed to be erratically tended, however, and the hydrangea was rangy and unkempt, flopping over the box hedge and obscuring the clean lines of the design. In an adjacent bunker compartment, an apparently self-seeded thicket of pear and alder rose over the concrete walls. Certain species seemed to emphasize the blurry boundary between cultivation and chaos most acutely. Across the site, silver birch trees grew in incongruous places—in inch-wide gaps between concrete blocks, along elevated rail lines, clinging to narrow ledges. But they had also seeded in groves in the interstitial open spaces on the site, where they appeared to have been allowed to mature without any interference—for the most part. In one of the sunken bunkers, a gap in the wall revealed a glimpse of a spiral of short birch logs, laid onto the dipped surface of the ground and slowly rotting. I later learned that the birch had been cut down in the early "vegetation management" of the derelict rail lines, and rather than being chipped or disposed of off-site, the wood had been repurposed to provide an object lesson in decay and regeneration, intention and intervention.

As I moved farther from the center of the site, I came across another strange zone, this one enclosed within a high brick wall. Inside, in the midst of a remnant orchard, a grid of squared beds and paths had been laid out, separated by sections of steel that resembled railroad ties. Each squared section had been filled with salvaged material composites— construction rubble, gravel, used bricks, broken tiles. The plants growing in these scrap substrates were familiar, opportunistic waste ground species: dock, bramble, willowherb, mullein. Two giant burdock plants leaned from their enclosures across a path to meet in the middle. But adjacent to these odd compartments there were also cultivated flower beds, thick with asters, daisies, and ornamental grasses.

I realized that I'd read about this space in a chapter by Elissa Rosenberg, where she describes these beds as "research plots": "True experimentation is defined by its open-ended quality, and the unpredictable results that often challenge the initial hypothesis. Here, the aesthetic of experimentation assumes, as its starting point, the dynamic flux of natural process."[24] Rosenberg points out that the garden's apparent embrace of disturbance and flux signals an intellectual alignment with postequilibrium ecological theory, in which "succession is a highly probabilistic and contingent process," not a steady progression to a stable state. As with other areas on the site, however, some intervention seemed to be required.

The test plots were suspended in a state of calculated barren-
ness, which continually rewound the successional clock to
the point at which the colonization of the industrial waste
by volunteer plant species would be most apparent. Given a
few years of nonintervention, the entire walled garden would
have been an impenetrable thicket of burdock and bramble.

Leaving the walled gardens and heading back toward the
complex of bunkers and furnaces, I came across another heap
of rubble, in composition almost identical to the deposit in
of one of the test plots. The heap had apparently been quite
recently dumped, and coarse construction waste was interspersed with scraps of rubbish, paper, and broken glass. The
material tumbled down a slope into one of the untended
bunkers, around thickets of established willow and buddleia.
I stopped and examined it. For some reason, this heap seemed
anomalous and intrusive, as if a hurried worker had dumped
it as a temporary fix but planned to return later to remove
it. I found myself wondering why this deposit seemed to
be "matter out of place" when the almost identical heap of
rubble I had just encountered in the walled garden seemed
somehow sanctioned and deliberate. Uncertainty implicated
me in making sense and forced me to question my own as-
sumptions about the invisible hand of the gardener, as well as
the slippery line between waste and wanted matter.

As I moved back toward the bulky steel and concrete forms
of the bunkers, furnaces, and rail lines I found myself puz-
zling over how to integrate my experience of the apparently
(if unevenly) cultivated spaces into my interpretation of the
management of the site's built structures, and my interest in
the ways in which they were, or were not, being allowed to actively decay and disintegrate. I realized that part of the trouble
came with the distinction I was used to making between architecture (as discrete and bounded) and environment (as

dispersed and distributed). This site was operating through a more porous principle. I recalled a comment made by Tilman Latz in the publication that documented NVA's Venice debate:

> We should maybe overcome the notion of a building, as this always asks for function. We could rather think of it as a landscape, an open system that can transform over time and assimilate as many functions and interpretations as possible. Then a strategy of "controlled decay and growth" is maybe applicable.[25]

A further clue to the site's underlying philosophy was revealed when I came across an unfamiliar term on an interpretation board near a patchy meadow, in German with no English translation: *Industrienatur*. I returned to the row of translated signs flanking the main entrance and found the following definition:

> The term "industrial nature" seems at first to be self-contradictory. The ravages of heavy industrial production utterly changed much of the natural landscape in the Ruhrgebeit. Fields, meadows and farmland were replaced by colleries, coal tips, steelworks and waste tips. But even during the era of industrial production a few plant and animal species managed to take root here and there. When the industrial sites became redundant nature swiftly repossessed the derelict areas. We have given the name "industrial nature" to the particular mixture of plant and animal species which have managed to adapt to what were in places very arduous living conditions. Here you can discover rare species, a colourful mass of blossoms and bizarre forms of growth. What makes these disused sites so particularly beautiful and fascinating is the singular

and unfamiliar link between derelict industrial sites and
untamed nature.[26]

The sign also included a map of nineteen different sites in
the Ruhrgebiet included in a *Route Industrienatur*, an "industrial nature trail" that supplemented the cultural and
historical emphasis of the *Route Industriekultur* and "allowed
visitors to learn more about very special and unique natural
phenomena."[27]

The passage above appeared to be the only information
available in English, although I realized that many of the German signs I'd encountered on the site must have been part of
the guided trail. The concept itself, as articulated above, was
a curious one. Nature is framed as other—something external to culture, capable of acts of repossession and untamed
intervention—but also as inextricably intertwined with industrial and cultural processes, to the point where any essential naturalness dissolves. The concept of *Industrienatur*
aligns in a sense with current ecological thinking about novel
ecosystems, which are defined as systems "of abiotic, biotic
and social components (and their interactions) that, by virtue
of human influence, differ from those that prevailed historically, having a tendency to self-organise and manifest novel
qualities without intensive human management."[28] The
double inflection is present here too, however. The artificial
ecosystem is produced by human interference and paradoxically sustained by human indifference.

At Duisburg, adoption of the concept of *Industrienatur*
as an interpretive framework seems to have allowed for the
cultural valuation of ecological components that in other
contexts would be considered undesirable. Walking back
through the site (now with the assistance of my German–
English dictionary), I noticed several signs that explained the

way that specific soils and microclimates support specialist plant and animals species, adapted to cope with extreme conditions, nutrient deficiency, and residual contamination.[29] One sign noted that biting stonecrop (*Sedum acre*) prefers highly compacted soils composed primarily of slag created during the smelting process. Another singled out the neophyte goldenrod (*Solidago canadiensis*), a North American immigrant that I recognized from the scrubby New England fields of my childhood. Interpretation highlighted the fact that certain plants were able to thrive on the site not in spite of but because of the way industrial process had transformed the landscape. Species that would have been classified as invasive in other contexts were distinguished through their association with the site's industrial past in an inverted value system that gave precedence to the refugees and invaders, the pioneers of a postpastoral nature.[30]

In this site, nature, as Rosenberg puts it, is "inextricably bound up with technology and shaped by social relationships and cultural memory."[31] It is possible to draw a connection here between Duisburg Nord and the (admittedly less redemptive) "technonatures" that Shiloh Krupar identifies at postmilitary sites in the American West, which emerge through a "complex and dynamic co-mingling of waste and nature, the social and ecology."[32] The tension between a framing of nature as an inherently compromised human artifact and nature as autonomous agent is central and irresolvable in such places. Materials—be they plants, steel, soil, or seeds— are only ever partly cultured, and management embodies the lived contradictions of "simultaneously taking charge and ceding control."[33] During my self-guided tour of Duisburg Nord, I encountered an elevated meadow of blooming wild carrot, thistle, and salsify, perched on the end of the rail line that once provided access to the ore bunkers from above.

The meadow was in its high summer glory, abundant and sovereign. When I returned the next day, the meadow had been mown down, and workers were spreading fine gravel on the shorn ground. It had been "cultured" again to create a staging area for a mountain bike race. This pragmatic disturbance of a paradoxical ecology might be best described as entropic gardening, a variant of the "rambunctious gardening" that frames Emma Marris's discussion of novel and emerging ecosystems.[34]

It is not entirely clear to what extent the embrace of *Industrienatur* is applied to the built structures at Duisburg, although in principle the concept would seem to encourage the blurring of the distinction between architecture and environment that I noted above. Duisburg's core built area appears to be selectively maintained, with some remnants allowed to decay and others held together through minimal repair. Much of the critical literature about the accommodation of ruination at other sites in the Ruhr seems to emphasize the underlying artifice at work. Dan Swanton writes of the approach adopted in the redevelopment of a steel plant at nearby Dortmund:

> The invasion of plant and animal life is managed so that
> the ruin maintains an air of authenticity, without threat-
> ening the physical presence of the blast furnace as vast
> artefact. The chemical and biological lives of these material
> remains have to be controlled, and ecological processes of
> disintegration and regeneration truncated for these ruins
> to perform.[35]

Torgeir Bangstad offers a similar critique in his work on Kokerei Hansa (Hansa Coking Plant), pointing out that the emphasis on contemporary ecology works to excuse past industrial excess and allows managers to operate in a space that appropriates the productivity of decay and regeneration without totally surrendering to it.[36] Other critics have suggested that there is a political and financial convenience to privileging passive renaturalization over active maintenance and repair, which can be allied with other manifestations of laissez-faire economic neoliberalism. Kerstin Barndt has also noted the way that the emphasis on ecological renewal functions to "leapfrog" from the preindustrial pastoral to the postindustrial future, eliding the difficult intervening period of military industrialism, when Ruhrgebiet production was of decisive importance in both world wars.[37] While all of these critiques are valid, there is still a tendency within this work to frame plant communities not as collectives of independent, agentive entities but as compliant pawns, mobilized to realize the aesthetic and interpretive objectives of designers and managers. They also seem to assume that continued maintenance of the illusion will require some attempt to preserve the structural integrity and stability of the remnant structures.

But it is also possible to imagine that at Duisburg there will be an accommodation of more profound collapse and transience, as well as a willingness to collaborate with plants and other agents to interpret shared histories and craft shared futures. This experiment is well underway, and those responsible for the site appear to be practicing a selective indifference to ongoing processes of decay and dissolution. When decay is actively accommodated, Simmel writes, "such indifference is, so to speak, a positive passivity, whereby man makes himself the accomplice of nature and of one of its inherent tendencies, which is dramatically opposed to his own interests."[38] Will process be allowed to run to the point where the structures at Duisburg are too derelict to access? Could access be managed cyclically, withdrawn for a period of time while structures are unsafe, then allowed again after a new stability has been reached through partial collapse? The site seems to want to leave these questions open, to force the visitor into uncomfortable encounters with potential risk and danger. (Will this rusty step hold my weight? How secure is that chunk of concrete?) While this may be a precarious and temporary state, it is a valuable one, in that it provides an alternative to the stability and stasis we are accustomed to experiencing in heritage spaces. The aesthetic and affective potency of site operates through a disclosure of the properties and propensities of matter in the process of its unmaking. To return to Simmel, "destruction here is not something senselessly coming from the outside but rather the realization of a tendency inherent in the deepest layer of existence of the destroyed."[39]

An attitude of "positive passivity" frames people as both "accomplices" (accommodating ongoing change) and antagonists (intervening selectively to check decay and deterioration). At Duisburg, there is no apparent attempt to resolve this tension. In the mode of accomplice, the human agent is destabilized and distributed, and the illusion of cultivation and control is abandoned to make space for a messier form of coexistence. This abandonment is not total, however, and the site does not apologize for its occasional reversion to more secure models of human subjectivity that rely on the disruption or arrest of process in the interests of material stability. In fact, much of the aesthetic potency of the site lies in the interstitial space between these modes of attention. Duisburg Nord operates through a principle of productive indeterminacy, in which the visitor is encouraged to experience (and negotiate) the tension between domesticated and wild, control and chaos, intervention and abandonment, process and preservation, intention and accident. This underlying ambiguity may be intentional and may include elements of artifice, but this too is perhaps inevitable. What the place succeeds in doing is foregrounding the impossibility of any critical clarity in the way that we approach such places. The site exposes itself as compromised and incomplete, and in this, at least, it succeeds. It is a space of possibility—possible unraveling as well as coming together—and in this, it offers one model for entropic heritage practice.[40]

Is it a transferable model, though? The conversations that have taken place between the architects of the plan for Duisburg Nord and those responsible for constructing a viable future for Kilmahew/St. Peter's would seem to suggest so. But there are other indications that the Duisburg model may be too anarchic for adoption at a site where the architectural legacy is more exceptional and external expectations are more constrained. Since 2012, a group of academics, architects, activists, and artists have gathered under the collective The Invisible College, working with NVA to gradually reoccupy Kilmahew/St. Peter's and carry out experimental research. The Invisible College website provides this description of the site and their work:

> The masterpiece of modernist architects Gillespie Kidd
> and Coia, St. Peter's Kilmahew opened in 1966 as a Roman
> Catholic seminary, only to be abandoned two decades later.
> Since then the building has fallen into spectacular ruin,
> celebrated by architectural pilgrims, vandals, ravers, dog
> walkers, and relic hunters alike. Despite being "A" listed, it
> has proved the wreck of numerous commercial schemes for
> restoration. But the history of the site is much older than
> the ruin. Kilmahew was originally an early Christian chapel,

then a castle, and in the nineteenth century an industrial-
ist's mansion, at the heart of a planted and managed estate.
Each historical phase has left behind relics, rhododendrons,
a walled garden, bridges, and burns.

The woods of Kilmahew are much more than a tragic
modernist ruin. They are the living document of a millennium, at least, of occupation, and, potentially, a unique
"laboratory" in which, through the mechanism of The
Invisible College, innovative ideas about the environment,
from restoring historic buildings to managing biodiversity,
may be speculated upon, experimented with, and tested.[41]

I was involved in an event hosted by The Invisible College
in September 2012, and I visited Kilmahew/St. Peter's again
in spring 2014, but my knowledge of the place relies largely
on the generosity of those with a deeper acquaintance with
the place.[42] In my discussion here, I focus on the aspect of
the place most relevant to the arguments I'm exploring in the
book, at the risk of underrepresenting some of the other conversations going on around the future of the site.

The potency of the seminary building complex in its
current state as a "tragic," "spectacular" modernist ruin is
clearly valued, and the aesthetic attraction of its extreme
dereliction is well documented. Photographs of the ruined
altar, overhung with charred beams, and the ravaged, vaulted
space of the former nave crowd Internet and print forums.
Among those debating the future of the site, there is a clear
desire to retain something of this elusive, dystopic character in any future plans. As I've already mentioned, the stated
intention "supports the principle of an *unfinished* work and
accepts a level of entropy."[43] Since the early involvement of
NVA, and through the Venice debate in 2010, there has been a
continual return to questions about the appropriate balance

between intervention and strategic neglect. Is it possible to conserve a ruin? Architectural historian Ed Hollis, in an essay prepared for the Biennale, implies that the rhetorical invocation of entropy is hollow if it does not involve an acceptance of unpredictable change. "If nothing happens St. Peter's will soon melt away, reduced by theft, arson, rhododendrons and rain."[44] He goes on to suggest that acceptance of unpredictability can only ever be partial, and uncomfortable. Citing Ruskin ("Watch an old building with anxious care; guard it as best you may . . . do not care about the unsightliness of the aid"), he suggests that the future for St. Peter's might be better understood as a continual negotiation over the inclusion of unsightly aids that allow for incremental reoccupation while preserving some element of indeterminacy and openness. The embedded contradiction in the plans for the site—the promise to "save this valuable resource from irreversible ruination" and to simultaneously embrace the free play of entropy—frames an impossible proposition.[45] Ruination, it appears, will be both retained and reversed. In this, I suppose, the parallel with Duisburg is clear. What is not clear is how this will play out on the ground. In the rest of this chapter, I try to imagine what it would look like to apply some of the principles and practices adopted at Duisburg on the grounds of a very different site. This is a speculative exercise, but it may be useful in flagging the limits, as well as the potential, in accepting "a level of entropy."

My first visit to Kilmahew/St. Peter's was in the context of an Invisible College workshop, and I joined a group of others in visiting the derelict seminary structures and exploring the grounds via an "audio drift."[46] During the visit, I was struck by a powerful sense of the challenges inherent in trying to have it both ways—ruin and reuse, process and preservation. It was also clear that philosophy was already coming up against practicality, even before any plans had been implemented. It had been determined that the obligation to make the seminary structures compliant with health and safety legislation would require the removal of the asbestos folded through the fabric of the structure, forcing an unstitching not by rain and time but by crowbar and power saw. The *Rhododendron ponticum*, which had spread unchecked across the site in the century and a half since its introduction as an ornamental, was now classed as a noxious invasive weed, ostensibly threatening the survival of vulnerable indigenous species. Meeting basic levels of ecological and environmental sanitation would, it seemed, require radical intervention at the site.

Before I returned for my second visit a couple of years later, I reviewed the master plan for the site, which had been prepared by landscape architecture firm erz Limited in 2011. I was particularly interested to tease out a sense of how specific aspects of the site's ruination would be addressed and to understand what fate was imagined for the plants that had occupied the site in the interval since abandonment. The plan was unequivocal about the need for some form of structured action: "All of the traces of human intervention, from medieval to modern, currently exist in a ruined state. These cultural fragments, alongside the unmanaged landscape, overtaken by invasive species, are in an advanced stage of decline."[47] The specter of "invasive species" was invoked from the outset, even within an expansive vision for reoccupation to proceed through a "generative," "metabolising" process of "incremental placemaking." "A state of permanent flux is integral to the concept," reads the plan, and the overarching principle for visitor experience is to retain "the sense of surprise and discovery that the site contains at present."[48] This all sounds good (and familiar, from the Duisburg model), but as the plan becomes more specific, some of the openness starts to close down. The "conservation philosophy" seeks to "accomplish an effective rescue of the buildings under consideration" through selective "repair, upgrade, intervention."[49] The plan expresses a desire to keep "the qualities of the current 'interim' state of particular locations" through selective retention of "regenerative" species.[50] Other species would not be so lucky: "Immediate intervention is required to remove the Rhododendron and other problem species [Japanese knotweed, Himalayan balsam] and to create a viable baseline position for new planting and management of the woodland and remnants of the designed landscape."[51] Most species subject to "vegetation management" remained unnamed. In the immediate setting of the seminary buildings, "Rank vegetation and debris [are] to be cleared, [and] original surfaces reinstated."[52] Apparently entropy was more attractive in the abstract than in its expression through the agency of individual plants.

When I visited Kilmahew again in May 2014, plans had moved on, and while no major work had yet taken place, both rhododendron and asbestos removal were scheduled to begin within the next few months. I visited the site in the company of three people who all knew it much better than I, and I'd decided before going that I'd give myself a discrete task to focus my attention. With the help of a 1967 edition of *The Observer's Book of Wildflowers* (published the year after the seminary opened), I would attempt to identify the plants that were growing on and around the ruined structures. As at Duisburg, I wanted to understand the role that specific ecological agents were playing in the making of the ruin, and to know, or try to know, the ecology as an antidote to seeing only the obvious aesthetic framing of the ruin as spectacle. An "Inventory of Trees, Shrubs & Plants in Kilmahew/St. Peter's Estate" had been carried out in 2013, but the seminary complex had not been included in the survey.

We entered the site on an unmarked path that dropped us into the woodland at the top of the estate near the remains of Kilmahew Castle. The castle aptly plays the role of the legible ruin on the site, almost to the point of absurdity. The central tower, dating roughly to the sixteenth century, rises in a clearing above the distant Clyde, its stone sides stippled with a picturesque veil of ivy. The structure resembles a traditional defended Scottish keep, but it is not clear whether it ever served this purpose.[53] When the fad for rusticated ruins reached this corner of Scotland, the owner of the estate decided to improve the structure's aesthetic with the insertion of some faux-Gothic windows, double ornamental corner niches, and a new columned entrance.[54] A fire in the late nineteenth century gutted the structure, which is now a shell, with saplings of goat willow and buddleia rooted inside on crumbling ledges. Perched on the hill above the ruined seminary, it seems to offer a mute object lesson in the folly of ruin improvement. In May, when we visited, the sham ruin was all wee leaf and filtered light. An owl nested up the old chimney, and the midden on the hearth was thick with rodent jawbones, bird skulls, and skeletal leaves. An analogue midden of dozens of Murray Mints wrappers occupied a gap in a nearby wall. The names of the plants clinging to the interior walls were an index of animal parts—hart's tongue, crane's bill, colt's foot.

We made our way down to the seminary along the course of one of the two burns that splice the estate, past thickets of rhododendron, around a small pond. At the seminary, we came up against a barricade: a high fence of steel slats, each sharpened to a threatening point. The fence had accreted layers of expired warning signs, most now illegible. The newest addition was a graphic of a stylized person whose lungs had been invaded by splotchy black spots: "Danger: Asbestos Dust Hazard. Cancer & lung disease hazard. Authorised personnel only." A second sign below read simply, "Dangerous building" with a yellow warning triangle. Across this someone had spray-painted, "LIES!!!" Inside, a sign posted by Reigart Contracts Ltd. ("Demolition. Recycling. Asbestos Removal"), read, "Danger: do not play in this area."

Our plant inventory took care not to disturb any of the building's lethal fabric. We began with the area around the cracked granite altar. The floor was boggy with rain that had fallen into the roofless structure, the puddles laced with a sour stew of human and leaf litter. Rank vegetation and debris rooted in the rotting concrete and in cracks in the walls. We started with a catalog of the ordinary army of first flank pioneers: dandelion, dock, bramble, nettle, plantain, willowherb, hawkweed, sow thistle, hart's tongue fern. We noted the odd specimen of maidenhair fern, forget-me-not, tutsan, suspected saxifrage. Several trees had taken root in the structure as well, their seeds cast on a strong winter wind perhaps, or carried in the fold of a raver's cuff: ash, willow, birch, Scots pine, hazel, Lawson cypress. One cypress specimen had its roots tangled in a snarl of red and white warning tape, but the tree's flat, feathery branches looked healthy and vigorous. The Lawson cypress is native to Oregon and California; it would have arrived on the estate at the behest of one of the nineteenth-century gardeners, who also introduced giant sequoia, California redwood, Douglas fir, and Himalayan cedar. Most of the trees in the ruined seminary were still saplings, less than head height, but the original specimens are now mature. One giant sequoia near the walled garden has swollen to a girth of almost seven meters.[55]

We continued our botanizing outside the main seminary building, on the foundation of the original Kilmahew House, which was destroyed by fire in the early 1990s and subsequently demolished. Within the footprint of the house, a dense thicket of birch trees has self-seeded, hiding the rusty hulk of an abandoned car. This is one of the only areas of spontaneous regeneration that will be retained, according to the master plan:

> The strategy for the spaces immediately associated with the seminary buildings is broadly focused on uncovering, repairing and reinstating the original finishes (hard and soft). . . . The counterpoint to this conventional conservation approach is the retention and management of the regenerative birch trees within the footprint of the former Kilmahew House. The retention of these trees acts as a marker of the interim derelict state of the site.[56]

The birch trees will be asked to function as a symbolic remnant of a discrete interval in the biography of the site, an interval that will be effectively closed by the restoration of the rest of the complex to its origins. The grove will function as a safe space for the evocation of entropy, if not its actual expression. "The dense block of birch trees echoes the original spatial arrangement, over time filling the physical void of the demolished house. Within this block of trees a discrete space is proposed for quiet, personal discovery."[57] Other patches of self-seeded birch will be retained in the tennis court and the walled garden.

The rest of it will go. As I write this, rhododendron removal has already begun, and asbestos removal won't be far behind. The odd assemblage of volunteer plants in the seminary structure will give way to a restored "original finish," and their stories will become untellable. For a moment, though, through our act of inventory, they came into focus not as a generic collective but as individuals—the spindly red stalks of the herb Robert, the familiar sleeve of the dock leaf, the scrappy saplings of ornamental conifers. They were not only a vegetative backdrop providing a screen of green to soften the gray concrete but also a community of volunteers—hopeful, patient, unaware of the landscape vision that would (eventually, probably) exterminate them. It is difficult to say whether there may have been an alternative to this excision, and exorcism. Has anyone tried to interpret the cultural history of asbestos in situ in a contaminated building, or through its removal? (Such an attempt could be framed, in Simmel's terms, as a "realization of a tendency inherent in the deepest layer of existence of the destroyed."[58]) How will the wholesale removal of the rhododendron alter the experience of this place, and how long will it take for the landscape to heal over the scars? What becomes painfully clear is that the unmaking of this site will be accomplished as much by people as by agents of chemical and ecological intervention, and that in the end, the free play of entropic process will be confined to very select playpens.

The potential for Kilmahew/St. Peter's lies, perhaps, in the willingness of those involved in determining its future to expose, rather than obscure, the compromises and contradictions that will trouble any reoccupation of this site. The ruin will be ruined, but exactly how it will be ruined is still an open question. At the 2012 Invisible College event, one of the speakers commented, "You can't tame a ruin. . . . The moment you make a perceptible constructive intervention . . . you have destroyed a large part of what gave the ruin its value."[59] Given the inevitability of the destruction of the ruin in its current state, the project turns to a framing of negotiation and debate as an end itself, with a continual looping back through the physical structure of the site to make incremental changes and craft provisional solutions. But can you have a master plan and still leave a site unmastered, open to different appropriations and inscriptions? At Duisburg, this seems to have been achieved, if precariously. The pressures at Kilmahew/St. Peter's are different, though, and the stakes are higher. In the end, the hard truth is that the vision for this site relies on securing support from funders and partners who may not be content with reassurances about openness, inclusivity, and improvisation. The future for this site will involve an opening up, most likely, but this will also mean a closing down of some of the possibilities that the site now holds, fraught with failure and collapse, scavenged and scarred, but still occupied, and appreciated.

At the close of this chapter I'm left feeling slightly dissatisfied with my conclusions. Is all that we can hope for honesty about our ambivalence, and the contradictions that riddle any attempt to collaborate with ecological and chemical process? I keep thinking about an example from much closer to home. I live in a terraced house in a small Cornish village, and our narrow backyard is hemmed in by neighbors' gardens on three sides. We have lived in the house for four years, and this summer the benign neglect of our borders came to a gentle sort of crisis, with the neighbors on all three sides requesting that we attend to the overgrown vegetation rooted on our side of the boundary line. We pulled stubborn willowherb from a high stone wall, hacked back sprawling skeins of bramble, and uprooted a buddleia that had tunneled under a lattice fence and was lifting the whole fence alarmingly in gusts of high wind. We brought our neighborly relations back into some kind of balance by checking the growth of the plants that we shared our space with. The ethic was pragmatic rather than collaborative. So to suggest that somehow it could be otherwise in places that are much larger and more complex, places that have a responsibility to a wider public, seems somehow disingenuous. But it is still something that I think is worth contemplating, if only because experiments like those taking place at Duisburg and at Kilmahew/St. Peter's allow us to imagine a different trajectory in relation to remnant ecologies and residual architectures, in which the words "restoration" and "preservation" sit awkwardly. Both of these words privilege recovery of an imagined original state over the discovery of possible future states—which are still rooted in the substance and stories carried forward from the past. And while attempts to reverse the needle may always involve a wavering commitment, they are worth exploring even in their wobble. Perhaps that is the promise of a positive passivity: the willingness to live with the wobble.

6

Boundary Work

ON EXPERTISE AND AMBIGUITY

> Before I built a wall I'd ask to know
> What I was walling in or walling out,
> And to whom I was like to give offense.
> Something there is that doesn't love a wall,
> That wants it down.
>
> *Robert Frost, "Mending Wall"*

WHEN I WAS A CHILD, I lived on a farm in Vermont, at the edge of a small town in the scoop of a valley. Our back fields butted up against the town cemetery, and its terraced lawns were part of our extended territory. One day I left the others behind and climbed the hill up to the far edge of the cemetery, where the gravestones thinned out and the encroaching woods cast an acid blanket of pine needles over the mown grass. I spied a small structure up there, on the far side of the slumped perimeter fence. I remember it now as a tiny house, though it was probably only a storage shed built to save the groundskeeper the long trip back down to the bottom of the hill. One corner of the roof had caved in, and the building seemed to have begun to turn inside out. Moss lay matted on the shelves, and drifts of leaves had gathered in the corners. A sapling sprouted from the rotten floorboards. But below the intact section of roof there was another shelf, and on this shelf lay a chipped saucer, a few rusted cans, and a fork.

I squeezed in around the door, which hung askew on one high hinge. Inside, I dusted off the shelf with a pine-needle broom and neatly arranged the saucer and fork. I scooped some of the leaves out of the corners and peeled the moss mats off the other shelves. My homemaking was play, but it was also quite serious. I was putting things in their place, and in doing so, I was reestablishing boundaries that had become frayed and porous: inside/outside; made/grown. I didn't spend long in the house that day, and I remember when I left the satisfaction of having arranged the world as it should be was cut with an undertow of another feeling that I didn't recognize. I don't think I ever went back to the house, and when I thought about it afterward, I always pictured it as it was when I found it, before I entered.

We are meddlers born. In this chapter, I talk about places that have something in common with the tiny house, places that have been left to their own devices for long enough to have begun to shed some of their original form and function. In these places, unchecked entropic process has produced hybrid assemblages where materials exchange and intermingle, and where processes of transformation are not explicitly contained or controlled. To return to J. B. Jackson, in these places the "interval of neglect" has extended for long enough that remnant structures and features have begun to slip over the threshold and shed their value as purely cultural artifacts.[1]

In each of the places I write about, however, there has also been a moment of recognition and revaluation. As soon as these sites become present as objects of concern, their mixed and mutable identity is framed as a problem to be solved. And the response, most often, involves a process of segregation and separation. With great effort, the tangles are untangled through a kind of semiotic and material boundary work. This boundary work (re)asserts the distinction between cultural and natural substance, and (most often) describes this work as an act of "preservation." Cindi Katz, writing about the management of natural environments, makes an argument that applies equally, in an inverted form, to the preservation of built heritage:

> Preservation represents an attempt both to delineate and maintain a boundary in space and to arrest time in the interests of a supposedly pristine nature which, of course, is neither bounded nor static. As such, preservation is quite unecological, defying natural history and the vibrancy of borders—physical, temporal, spatial—where evolution, change, and challenge are negotiated and worked out in nature as in culture.[2]

In order to consolidate the meaning, and the value, of the object of concern (be it a rare species or a significant structure), appropriate boundaries must be maintained (or reinstated). Various practices are involved in the labor of boundary maintenance: repair, restoration, rehabilitation. All of these practices establish as their goal a desired state of being (as measured by the presence of target species, or the integrity of the original fabric of a structure). The ultimate aim is framed as averting loss, and the greater the evidence of cross-contamination—through incursions of undesirable culture into natural habitats, or through the degradation of cultural artifacts by environmental process—the greater the assumed risk.[3] The boundaries that are maintained through these processes of demarcation and designation are not only physical, however; they are also professional. Those who are trained in a specific area of expertise—be it nature conservation or archaeology—establish their authority in part by asserting the distinction between their body of expert knowledge and that held by other disciplines. As Sara B. Pritchard has observed, "Analysing expertise . . . often opens up contestation over nature [or culture]: what it is, . . . how it should be managed, and by whom."[4]

In some heritage contexts, the rigidity of professional boundaries and the adherence to the preservation paradigm are giving way to recognition that processes of change and creative transformation can be productive and positive, and that the assumed boundaries between cultural and natural heritage are increasingly irrelevant and unhelpful.[5] Those who work in the National Trust are fond of quoting a 1996 report written by Alan Holland and Kate Rawles, in which the authors argue that "conservation is about managing the transition from past to future in such a way as to secure the transfer of maximum significance."[6] Such management, Holland and Rawles point out, requires the continual "negotiation between cultural and natural imperatives."[7] These sentiments are echoed by Gustavo F. Araoz, president of the International Council on Monuments and Sites, who has argued that cultural heritage practitioners should think about their role as managing change in an effort to protect value and significance rather than preventing change in a struggle to preserve the existing material heritage.[8] There are interesting parallels in the field of ecological restoration, where some scholars have also begun to question the way that historic reference conditions are used to guide contemporary management: "If natural states are elusive, if the environment is always changing and ecosystems are always coming and going, and if multiple realizations are normal, then the premises underlying ecological restoration to a historic standard come into question." Restoration, the argument follows, should be framed as "managing for change, which is accepted as inevitable."[9]

But old habits die hard, and while there may be a growing willingness to value change rather than stasis, there are still clear limits to acceptable change. It is remarkably difficult to find examples of conjoined natural and cultural heritage management that do not resort to some form of categorical fixing. The full realization of such a heritage practice would require what Rodney Harrison has described as an "ontological politics of connectivity," in which human and other-than-human actors form an emergent collective, oriented toward *becoming* rather than *being*.[10] Managers may be open to dynamism and change in principle, but this openness often closes down when faced with the prospect of border crossing and ontological mixing. To be fair, this reversion to type is often forced by the need to comply with regulatory frameworks, which stipulate that designated species or structures must be protected and their individual integrity maintained. The tension also partly hinges on issues of scale. While there is a broad recognition that landscapes are inherently hybrid natural–cultural entities, dynamic and ever changing, this recognition is rarely applied to the individual parts (bogs, barns, hedges, houses) that make up a landscape.[11] It is much more common to come across partial attempts at (habitat) restoration alongside selective (structural) stabilization, site-specific acts of boundary work that reassert the old orders.

In this chapter, I write about three places where an interval of benign neglect has given way to a phase of more active management: an engine house complex in the middle of Cornwall; a former gunpowder works (now managed as a nature reserve); and an abandoned Montana gold mining camp. Each of these places demonstrates the application of some form of boundary work. In one place, this work is abrupt and invasive, introducing a sharp discontinuity to a landscape previously characterized by gradual change and transformation. In another, the intervention is more tentative and compromised, with attempts made to mask the effect of interference. At the final site I consider in this chapter, the boundary work occurs primarily through an overlay of interpretation rather than through physical modification.

I've lived in Cornwall for eight years now, and its landscape still perplexes me. Growing up in New England, I was accustomed to discovering the traces of former industry and inhabitation in the hilly second- or third-growth forests—mossy spines of stone walls that once edged cleared fields, sunken cellar holes, oxidized farm implements, feral orchards. In Cornwall the remnants of prior occupation are of another order altogether. In the most remote locations, on an inaccessible upland moor, or along a thickly wooded riverbank, one frequently comes across the ruins of massive masonry structures—wheel pits and processing works, engine houses and auxiliary buildings, the residues of intensive mining, quarrying, milling, and manufacture. Cornwall's industrialization was precocious and, ultimately, precarious. Many of the sites one encounters have been abandoned for a century or more. There has been plenty of time for disused structure and infrastructure to be worked on by weather and weeds, gravity and gradual decay.

In the long interval of neglect, new ecologies sprang up in and around many of the abandoned sites.[12] Rare mosses and lichens established themselves in the copper-rich soil and on spoil heaps. Gorse and willow scrub rooted in the scarred open ground. Bats came to roost in vacant mine shafts and adits. Thick mats of ivy engulfed the walls of derelict structures. Birds nested in cavities in crumbling walls. Wild strawberries colonized seams in old concrete, and brambles clambered everywhere. For decades, these places were more or less ignored and overlooked—unless their building stone was needed for a project elsewhere, in which case they were quarried in situ. Beginning in the 1980s, some of these postindustrial sites achieved a state of ecological exceptionalism that justified their designation as a Site of Special Scientific Interest or a Special Area of Conservation; others were incorporated into nature reserves.[13] At about the same time, many of the structural remains within these landscape contexts were also listed as Sites of Special Architectural or Historic Interest, but there were limited resources available for conservation work.[14] Interest, in this instance, did not necessarily mean intervention, and as long as these places remained essentially unmanaged, the overlay of values was not a cause for concern.

UNESCO's designation of the Cornish Mining World Heritage Site in 2006 catalyzed a subtle but undeniable shift in priorities. The recognition of the "Outstanding Universal Value" of the remains associated with Cornwall's industrial past led to a drive to conserve and consolidate structures that were now cast as threatened and vulnerable, at risk of irreversible degradation. Resources began to flow into the county to carry out the necessary work, and places that had been gently neglected for decades became construction sites, then tidy facsimiles of their former selves, encased in smug new skins of repointed and gleaming granite.

In June 2014 I came across a news item in the Web archive of the local paper. The article, which had been posted a few months earlier, is titled, "Conserving Mining History to Commence at Wheal Busy."[15] The article mentioned that the Wheal Busy complex, near the village of Chacewater, was to be the next site to benefit from conservation funding that had been secured to stabilize threatened features within the World Heritage Site.[16] During its lifetime, the mine had produced substantial quantities of tin, copper, and arsenic, and its own-ers were early adopters of innovations in steam technology. The article mentioned that initial conservation work would focus on an 1856 engine house and chimney, disused since mining had ceased in the early part of the twentieth century. It also noted that the work included commissioning of eco-logical surveys to "assess the nature of the flora and fauna on the site, with particular regard for bryophyte species and bats" and to ensure that building conservation work would be done with "due regard for the site's varied ecology."[17]

At the end of the article, a man who identified himself as Will Rowland had posted this comment:

> Ecological Survey!!?? It's a joke surely!!! . . . so far Natural
> England and Tregothnan Estate have done an ABYSMAL
> job of conserving the Ecology at Wheal Busy Tin Mine/
> Ecology Garden. . . . Protected species displaced/destroyed
> so far[:] Barn owls . . . were nesting in the Engine House,
> Southern Marsh Orchids bulldozed up at rear of mine,
> breeding populations of Grass Snakes . . . Palmate newts
> (some an acquaintance managed to rescue and move to an-
> other pond after I tipped him off.) The Ivy has been indis-
> criminately butchered at the base and ripped down at the
> rear of the mine. I am starting to think that the "Ecological
> surveyors" at Natural England spent most of their time
> down the Kings Head Pub, Chacewater! The vast area of ivy
> provided a fantastic habitat for numerous invertebrates and
> the berries provided a feast in late autumn and winter for
> Redwings, Blackbirds, wood pigeons etc. Goldcrests nested
> above the arches. I have a friend doing a PROPER bat sur-
> vey of the mine soon, as Natural England either don't seem
> to have a clue . . . or are getting paid off by someone! The
> Ivy should have been removed in stages over a few years

and only cut halfway down on some walls. This would have
allowed for re-pointing of the walls in stages. And as for
any rare bryophytes mentioned in the article . . . Now well
and truly BULLDOZED!!![18]

Intrigued, I engineered a visit to Wheal Busy in early June,
where I parked in an unpaved, puddled lot next to a faded
information panel. The sign mentioned the historic value of
the mining industry remains, but it also referred to the area's
unique flora and fauna, most notably greater horseshoe bats
and rare liverworts. I crossed the road and walked past a large
one-story stone structure that was built as the mine's smithy,
now apparently used for storage. As I rounded the corner of
the building, the Wheal Busy engine house came into view,
rising above a tangle of young trees and shrubs, crowned and
draped with a bristling thicket of mature ivy. As I drew closer,
I could see the fresh tracks in the ground around the structure, raw seeping scrapes and flattened vegetation. At the
back of the engine house, near the chimney, the ivy's wrist-
thick vines had been severed at the ground. There was no
evidence of other intervention, aside from a few posts and
pots that I took to be part of the "ecology garden" mentioned
in the irate post.

   When I returned again four months later, the site was
cordoned off with steel fencing, and the work was well underway. The engine house and its chimney were sheathed in
scaffolding, and the area around the structure was bare and
compacted. Heaps of stripped ivy mounded up at the edge
of the site in waste piles. The work on the engine house appeared to be almost finished—surfaces scraped clean, seams
repointed, exposed courses capped with concrete. The admixture of decay and regrowth—Riegl's age value, or Ruskin's
patina—had been replaced with a roofless replica of the

structure that would have stood on the site, brand new, in 1856. Time had been selectively rewound, the engine house scoured and saved and firmly established as an object of heritage. In the process of being made present as heritage, the structure had been effectively de-natured, and its ancillary ecologies had receded to the point of irrelevance, despite a stated intention to proceed with "due regard." The restored structure was curiously blank; it communicated little to me beyond its signification of the iconic engine house form: a fine specimen, but not a living one. The boundary work was almost complete. Jackdaws circled the scaffolding, impatient to take up residence again.

A few months later, I had an opportunity to discuss the work that had been carried out at Wheal Busy with a Natural England land manager. He wasn't familiar with the specific details of the work, but he pointed out that the contractors who did the consolidation of the engine house would have had to look out for potentially adverse impact on any "species of significance" while they were doing the job.[19] (During another recent engine house restoration, when they realized that the cracks and crevasses in the structure were being used by breeding birds, they waited until the young had fledged, then preserved the nest holes with wedges of foam while they were mortaring the rest of the structure.) For the most part, however, the ecologists would have deferred authority to the archaeologists and contractors responsible for carrying out the work. The boundary between the different areas of expertise would be maintained for as long as it took for the structure to be reassembled—and then the process of renaturalization would begin again.

A few miles away from Wheal Busy lies another complex of structures associated with Cornwall's mining history. The gunpowder works at Kennall Vale were established in 1812 to produce explosives, primarily for industrial use. Here, in a steep-sided, thickly wooded valley, manufacture harnessed the hydropower of the Kennall River to process saltpeter, sulfur, charcoal, and graphite into a (relatively) stable product. Mines used the gunpowder to extract ore from below-ground shafts, and quarries used it to loosen massive blocks of stone. Mistakes were sometimes made (one accident was caused by a hot roasted potato in the press house), and several workers lost their lives in accidental explosions. Production declined in the early part of the twentieth century with the development of new nitroglycerin-based explosive technologies, demand accelerated by the onset of World War I. Back in Kennall Vale, at the edge of the disused gunpowder works, a granite quarry was established, specializing in fine-grained granite for use in memorials to fallen soldiers.

Now, over a century since the closure of the works, the landscape holds little to alert the casual visitor to the intensity and the volatility of its past life. When my family first moved to Cornwall, we lived only a couple of miles away from Kennall Vale, and we used to visit often. We walked the paths cut along the side of the valley, following an elaborate system of leats that once directed waterpower to a series of incorporating mills. The mills and other industrial structures were now overgrown with ivy and ferns, picturesque relics that attracted local photographers to document their dereliction. Bluebells carpeted the forest floor in the spring, up to the base of the massive blast wall at the top of the valley, which had been built to protect adjacent properties. On the side of the valley across from the mills, the granite quarry void had filled to a still, shaded pool. Mosses encased blocks of waste granite, forming soft green cubes.

Everywhere there were trees. Mature beech, sycamore,

ash, and oak grew densely through the valley and rooted in the halfling structures. My favorite specimen was a young sycamore that had grown up next to a blocky concrete foundation. Nudged by the wind against a sharp corner in the structure as it grew, it had crafted an ingenious solution by molding its trunk with a matching right angle, which now rested on the concrete ledge like an elbow propped on a wall. Many of the older trees were deliberately planted by the gunpowder works employees to keep humidity levels high in the valley. The hope was that their presence would reduce the likelihood of accidental detonations and contain them if they did occur—a sort of green blast chamber.[20] With the withdrawal of active industry, the trees accelerated their occupation of the landscape. By 1985, the valley was sufficiently renaturalized to merit its designation as a nature reserve, managed by the Cornwall Wildlife Trust. The derelict walls and structures provided habitat for pioneer plant species, and greater horseshoe bats roosted in nearby mine shafts. Kingfishers and song thrushes were spotted in the valley, trout and eels in the river and leats.

The first historic and archaeological survey of the site was completed in 1986, a year after the designation of the reserve.[21] The historic value of the site was deemed to be significant enough for the remains of the gunpowder works to be designated as a Scheduled Monument in 1999. Management of the structures up to this point had been minimal, but an archaeological assessment prepared in the same year as the listing identified some potential issues:

> The effects of mature trees falling onto structures or damaging them as roots are lifted is a serious concern. Some leats have already been affected by falling trees. Saplings have also begun to establish themselves on and in the walls

of some buildings. Dealing with the latter should cause no
problems to the CWT [Cornwall Wildlife Trust] but the
extensive and awkward tree surgery needed to ensure that
important remains are not damaged by falling trees will not
only be expensive but may also lead to clashes with nature
conservation aims, not least the retention of the large
mature trees and their communities of invertebrates, bryophytes, lichens, fungi, etc. It should also be expected that
the opening up of numerous small glades around buildings
will probably lead to the development of scrub woodland
whose root systems may also have a negative impact on the
archaeological remains and will certainly render them less
easily visible. The historic character of the reserve, in which
the trees were themselves used as devices to minimise the
impact of any explosions in the gunpowder works, would
also be altered by the fragmentation of the woodland.[22]

In the passage above, the trees are cast in at least three distinct
roles: they are ecological entities, offering essential habitat for
specific species and providing structure within the broader
ecosystem; they are unintentional agents of destruction,
threatening to undermine buildings and obscure archaeological remains; and they are cultural artifacts in their own right,
their presence contributing to the historic character of the
reserve by signaling their past enrollment in the industrial
process. This passage was written a quarter of a century ago,
and, as my recent exploration of the site suggests, the trees
are still there, and their identity remains unfixed. If anything,
they have accrued another part in the performance, as the
nonnative beech and sycamore (planted during the period of
active industry) are occasionally demoted to second-class citi-
zenship in favor of indigenous oak and ash.[23] At Kennall Vale,
nature conservation value and cultural heritage value seem to

be treated with a rough equivalence. The boundary work that does take place is, for the most part, selective, and focused on specific interventions rather than wholesale restoration. The 1999 report recommended removing ivy and saplings from the structures, and over the years there has been a fair amount of this kind of remedial maintenance, but much of the vegetation has grown back over time.

In 2009, a spike of anxiety about the gradual deterioration of the structures led a group of volunteers and historic environment professionals to carry out a photographic survey of all buildings and structures within the reserve. Scrub and saplings were removed from around the structures to make them visible for the photographs. The survey identified vandalism (local youths were fond of taking loose stones from the buildings and tossing them into the river) and "tree-throw and tree-root damage" (a variation on vandalism, but with arboreal rather than human agents of destruction) as the key factors undermining the condition of structures. Trees were considered to pose "major threats" to public safety and to the historic buildings.[24] A flurry of activity followed over the next several years, some of the work funded by the same source that had financed the work at Wheal Busy. Across the site, ostensibly overmature trees were felled, and selected buildings were reclaimed and repointed.

Those who carried out the work were aware that structures without a protective covering of ivy and moss were more likely to be the target of vandals. In removing the vegetation, they were stabilizing the structures but simultaneously increasing their vulnerability. On some buildings, new mortar was covered with earth to encourage moss and lichen to recolonize—an ecological and aesthetic strategy of camouflage. In other areas, vine and root had come to provide structural support, taking the place of decayed mortar, and reconstruction involved difficult decisions. A set of steps next to one of the mill buildings had been assembled using recycled millstones, but because "the roots of a large tree [were] intimately wound around the steps, with one root even forming one of the steps," the work they could do was limited.[25] The willingness to negotiate, or even collaborate, with vegetative agents has its parallels in the practice of "soft capping" ruined structures with layers of protective, living matter.[26]

The language in recent reports and the management plans for Kennall Vale retains an element of ambiguity. A 2011 report prepared by English Heritage's senior field officer in Cornwall stresses the need to hold back the "ongoing tide of natural decay and vandalism," without acknowledging that decay is an essential process in any woodland ecosystem.[27] On the ground, however, managers attempt to find synergies between seemingly disparate goals. Recent repairs of the leat system, which involved the removal of several mature trees, also apparently provided a more consistent flow in the watercourses for plants, bryophytes, and wildlife. "Removing inappropriate species and improving the woodlands structure by selective felling, ride creation, restoring glades and rotational coppicing" has been found to create new pockets of habitat while simultaneously protecting the structures from the effects of unpredictable "tree-throw."[28] Although there is an air of convenient coincidence about the rationale put forward for these management actions, there is also a sense that the managers on this site are attuned to the way their labor must continually negotiate the nature–culture boundary. They accept that there will be compromises and that sometimes different aims will clash, but for the most part Kennall Vale seems held in a precarious balance. A visit to the valley can still offer an

illusion of nonintervention, the structures surrendered to a
state of pleasing decay, surrounded by semiwild nature.

In the United Kingdom, a place like Kennall Vale—valued
for elements of both natural and cultural significance—is
unremarkable. Designated National Nature Reserves (sites
deemed worthy of a higher level of protection than Kennall
Vale) include within their borders hundreds of scheduled
monuments and other cultural resources.[29] The traces of
centuries of habitation and industry lie thickly on the landscape, and while boundary work is still occasionally carried
out to maintain the integrity of various components, there
is a general willingness to accept the ambiguity that arises
from inevitable intermingling. Features old enough to have
lost their form and structure (and so considered to be beyond
saving) are valued for their contribution to nature conservation goals: rare lichens colonize worked stone; common red-
starts and spotted flycatchers nest in gaps in ruined walls.

In the United States, the attitude toward such hybrid
landscapes is more anxious and unresolved. In 2004, during
negotiations over the boundaries of a new wilderness area
on Lake Superior's Apostle Islands, areas retaining material
evidence of intensive former industry (mostly farming and
quarrying) were largely excluded from the designation. James
Feldman writes, "The result is a wilderness boundary created
for management, one with boundaries that do not blur the
lines between nature and culture—a legible wilderness."[30]
In his study of the negotiations, Feldman observes that the
attempt to maintain clear boundaries between wild nature
and cultured landscape was largely a matter of interpretation
rather than strict adherence to legislative guidance. A broad
application of the 1964 Wilderness Act would have accommodated a designation that included the relics of former

farmsteads and the traces of industrial production, and that approached these features as artifacts of both natural and human history. Feldman argues for a new measure of "legibility," one that denies the segregation of nature and culture and makes visible the ways that "two seemingly distinct categories overlap" in rewilded landscapes.[31]

I was aware of Feldman's research when, in 2005, I had the opportunity to work as a cultural resources consultant in Coloma, Montana, an abandoned gold mining camp in the Garnet Mountains. The interval of industry at Coloma was fairly brief: mineral extraction began in 1895 (in its heyday, the camp's population supported several saloons, a boardinghouse, a post office, a school, and a reading room), but by 1910 the mines were failing and the residents had begun to disperse. The camp was inhabited up until 1960s by an intermittent population of hermits and hippies, and then in the 1970s it began to draw attention from treasure hunters, intrepid ghost town tourists, and historic preservationists. The federal Bureau of Land Management (BLM) owned a significant portion of the land in the former camp, which was cut through with private land holdings linked to former mining claims. The BLM carried out a cultural resource survey of the site in 1981, but because many of the remaining buildings were on private land, no stabilization work was initiated. A few miles down the road, the BLM made a considerable investment in the restoration of a contemporaneous mining camp, which became known as Garnet Ghost Town. At Garnet, dozens of cabins were stabilized and several commercial buildings reconstructed to give an overall impression of "arrested decay."[32] As Garnet was frozen in a state of suspended animation, Coloma melted. Structures and features continued to deteriorate, and cultural remains rapidly crossed over the threshold between history and archaeology.

Coloma perches on a broad shelf of land above the Black-foot River Valley at an elevation of almost 6,000 feet, and it is usually snowbound into April. When I first visited in July 2005, I drove in along a narrow track off the main road; a few hundred yards in, the thick stand of Douglas fir and lodge-pole pine opened up into to a wide, grassy slope punctuated by thickets of thimbleberry and solitary trees. At the edges of the open area, heaps of waste rock brooded, their bare slopes colonized by the occasional young conifer. As I walked through the site, it became clear that the thickets of thimble-berry were crowded around angular heaps of logs, the husks of former cabins. Toward the edge of the shelf of land, where the land fell away to the valley, a row of more intact cabins resembled giant games of pickup sticks, logs splayed and skewed by the weight of winter snows. Along the street where the commercial buildings had once been concentrated, there were a few granite rubble foundations, now overgrown with wild raspberry. A handful of buildings still retained sections of roofing. One small frame house was almost eerily intact. Looking closer, I saw a twisted bed frame in a collapsed cabin where scraps of newspaper on the walls documented ac-counts of Indian raids and land speculation. The scattered waste dumps appeared to have been well picked over and re-populated with contemporary rubbish. There was other evi-dence of people occupying the site—fire pits and fresh nails in the trees—but no sign of any intervention to stay the ef-fects of weathering and decay.

I had been hired by the BLM to help with drafting a man-agement plan for Coloma and to make some (belated) sugges-tions for the protection of its cultural resources after a recent property acquisition that included some of the land within the boundaries of the former camp. As a federal agency, the BLM has a legal obligation to manage the historic features that fall within in its extensive holdings, and while Coloma was not formally listed on the National Register of Historic Places, it had been declared eligible for such a listing. But the site presented a conundrum. Fewer than a dozen structures retained enough of their integrity to be viable candidates for remedial stabilization or repair. The others were too far gone, and essential information about their original forms had been erased through years of decay. In the absence of documentary and structural evidence, accurate repair and reconstruction was impossible. Yet information about the site's history was amply available in other records if you knew how to read them. Tree-ring evidence suggested that most of the trees that had provided logs for the construction of the cabins had been ninety to a hundred years old when they were cut, an interval that dated their germination to immediately after a major recorded wildfire event in 1805. The period of postindustrial abandonment, also spanning roughly a century, had produced a stand of lodgepole pine on the site of approximately the same size and age as the cabin logs. The trees at Coloma—as at Kennall Vale—were carriers of both cultural and ecological information.[33]

As I began to familiarize myself with the relevant policy guidance around the BLM's management of cultural resources, I came across an interesting entry in the BLM's manual for *Identifying and Evaluating Cultural Resources* that outlined the contexts within which application of an "experimental use" category might be appropriate:

> This category may be applied to a cultural property judged well-suited for controlled experimental study . . . which would result in the property's alteration, possibly including loss of integrity and destruction of physical elements. Committing cultural properties or the data they contain to

loss must be justified in terms of specific information that
would be gained and how it would aid in the management
of other cultural properties. Experimental study should
aim toward understanding the kind and rates of natural
or human-caused deterioration.[34]

Here, it seemed, was permission to develop a strategy for
Coloma that proposed doing things otherwise, sidestepping
some of the boundary work that would have required managers to reestablish the distinction between cultural and natural
resources. The "loss of integrity and destruction of physical
elements" could be justified if other forms of information
would be generated through the process. Management, in
this context, might involve close observation and monitoring
of gradual deterioration, rather than intervention to ensure
structural stability. The concept of experimental use reso-
nates with an idea developed by a friend of mine, a curator
at a prominent U.K. industrial heritage site, who thinks that
we need "heritage body farms" where terminal structures can
be subject to study analogous to that carried out by forensic
scientists on decaying corpses.[35]

At Coloma, the groundwork for such experimental use
had already been initiated by historical archaeologists from
the University of Montana, who were in the process of setting
up a platform for a virtual ghost town. They planned to use
laser scanning techniques to create 3-D digital models of all
of Coloma's remnant structures, then to stitch the scans together in a geographic information system representing the
wider landscape context.[36] As I began to develop the management plan, it became clear that the survey would also leave its
own record on the site: the student technicians hacked down
dozens of young trees to allow their laser scanners a clear line
of site to the structures. On the whole, however, the research

being carried out by the archaeologists seemed broadly compatible with an experimental use classification.

I wrote a first draft of the management plan outlining the proposed approach, calling attention to precedents elsewhere. I'd come across mention of a policy of "managed decay" that was being applied to the historic remains of the sealing industry on islands in the Australian Antarctic. The management plan for Heard Island and the MacDonald Islands Reserve acknowledged the "threat of loss of cultural heritage values" but sought to minimize, mitigate, or avoid adverse impacts rather than to stabilize or conserve. Images of the reserve showed elephant seals lolling around on the rusted remains of the iron drums that once boiled down the blubber of their ancestors for oil.[37] In the plan I wrote for Coloma, I tried to suggest that while efforts should be made to avoid unnecessary destruction and deterioration, where loss of integrity was unavoidable, there could be ways of interpreting change and transformation that deepened, rather than diminished, our understanding of the site's history. The draft proposed an interpretive strategy that treated the trees as living clocks, their gradual recolonization indexing change over time. Rebecca Solnit, in an essay about a rephotography project in Yosemite, also writes about trees as timekeepers, their rings "a calendar that suggests that time is cumulative, the present surrounds and contains the past."[38] More prosaically, the plan suggested a program be put in place to systematically monitor rates of decay, measured against a comprehensive baseline inventory of the condition of the structures.

When I shared the draft plan with the local archaeologist and other BLM managers, they requested that the proposal for an experimental use classification be moved to an appendix and that new language be introduced to emphasize the

BLM's commitment to protecting the remaining resources. The final draft of the plan made these changes, and I handed the work over to the BLM. The plan was never formally adopted, and I still don't know exactly why. The proposals in the plan have been inadvertently implemented, however, in that the BLM has not had the resources to carry out any stabilization, and the condition of the structures has continued to deteriorate. In this case the absence of intervention is classed as a failure rather than an opportunity, as (I think) it could have been. I'm perhaps too close to Coloma to provide an impartial account. Instead, all I can offer is a story about one place and its perplexities, and one attempt to work across the boundaries that construct (and constrain) heritage practice.

The work carried out in each of these places resembles my housekeeping in the ruined shed at the edge of the Vermont cemetery in some obvious ways. An initial attraction quickly transitions into a desire to intervene to reassert the clarity and definition of blurred boundaries. To return to Mary Douglas, in these places, ambiguity threatens to undermine the identity of the objects of concern. But what happens if we are willing to let identity adhere *in* ambiguity?

Katarina Saltzman has proposed that a "composting" metaphor might be useful in these contexts. If the fundamental principle of composting is that all "decomposing is simultaneously a process of composing," she suggests, we should be alert to the potential for "something new and useful" to arise when materials (and meanings) begin to break down and lose their integrity.[39] The process of composting is unpredictable and complex, in that it does not respect any of the categories that are usually applied to sort the world into perceptible parts. By shifting the focus of attention from "deterioration to compilation,"[40] Saltzman argues that a composting metaphor opens up radically new interpretations of remaindered things and places. When we begin to understand certain places as caught up in a process of composting, we can appreciate that dereliction and abandonment are not unfortunate end points (and problems to be solved with intervention) but rather transitional states, where different uses (social, ecological, cultural) combine to generate new possibilities. "The metaphor of composting can help us to better recognize and understand what happens in-between the before and after, to analyse quietly working microprocesses of decomposition and composition, and to provide alternative understandings of time and of mixing," she writes.[41] If one way to think about the ideas I've been exploring in this book is as a call for a postpreservation paradigm, then Saltzman's contribution is to think instead of a compostheritage practice that insists we are never "post" anything but that instead we carry our inherited (and invariably moldy) ideas with us as we move forward into new modes of action and attention.

There are a few interesting examples of instances when adoption of a compostheritage ethic has come about inadvertently, through attempts to reconcile species-level protection with built heritage conservation. All bat species in Britain are protected under U.K. and European legislation, and owners of properties where bats have taken up residence must not disturb their roosting places or the bats' persons. The result is often uneasy cohabitation, and while some property owners may welcome the presence of bats, others do not. In June 2014, Conservative peer Lord Cormack opened a debate in the House of Lords about the corrosive effect of bat droppings on the fabric of historic churches and on parishioner morale. At one point, Lord Cormack asserted, "If this debate achieved only one thing—a better balance between the demands of English Nature and the needs of English

Heritage—I would be well content."[42] A guide recently published by the National Trust sought to introduce "some of the
main dilemmas involved in sustainably managing wildlife for
builders, planners, architects and those of us lucky enough
to share our dwellings with wildlife."[43] The optimistic end
note is undercut by the candid acknowledgment of "main
dilemmas."

At some historic properties, the species of concern is not
animal but bryological. Thatch moss (*Leptodontium gemmascens*) grows only on decaying vegetation, and its preferred
habitat is decomposing roof thatch. The moss is listed as "vulnerable" in Britain and is classed as a Biodiversity Action Plan
species.[44] It has been found on Thomas Hardy's cottage in
Dorset and on the Holnicote and Killerton estates in Devon,
where the "main dilemma" involves balancing the needs of
the moss with the needs of the structure below. Managers
must walk a fine line between accommodating rot and repel-
ling rain, which means treating the building essentially as a
living organism.

As these examples demonstrate, boundary work is a continual chore for those tasked with "protecting" species and
structures of significance, and in the messy realm of every-
day practice, boundaries and borders are usually "less self-
evident, more unstable, and more multifaceted" than would
be suggested by reading of management plans and policy
guidance alone.[45] Recognition of this productive indeterminacy can only be achieved if there is a genuine move to, as
Steve Hinchliffe has suggested, "reconstitute nature conservation" (and reconstitute heritage preservation) in order to
make valued things (animals, plants, buildings) present without reverting to static, bounded conceptions of identity.[46]
In a 2005 paper about points of divergence and convergence
between natural and cultural heritage discourses, David

Lowenthal observes that both share a concern with the "protection" of "precious and irreplaceable resources" and seek to hold their inheritance "in trust for future generations."[47] While he draws a parallel between the sanctification of native species and the fetishization of the original fabric of a structure, he identifies the key difference in public attitudes to interference, which is seen as necessary for the protection of cultural artifacts and as something to be avoided in ostensibly natural environments.

As this chapter has shown, interference can take many forms, and its application is certainly not confined to the management of built heritage. Perhaps the future for an expanded heritage practice lies in a willingness to trouble the distinction between *making* and *growing*, and to accept the implications of a perspective in which all structures and artifacts have biological as well as social lives, and all ecosystems have deeply cultural properties as well as natural ones.[48] If this kind of shift were to happen, ambiguous identity may no longer be so threatening. We may find ourselves with a new ability to appreciate porous and perforated places, where interference is applied to highlight points of exchange and intermixing rather than to police the borders. In relation to the places I've been discussing in this chapter, such a practice may also allow us to draw critical attention to the historical specificity of obsolescence and abandonment, and to find fertile ground for interpretation among the old rot and the new roots.

Palliative Curation

THE DEATH OF A LIGHTHOUSE

> the art of losing's not too hard to master
> though it may look like (*Write* it!) like disaster.
> *Elizabeth Bishop, "One Art"*

THIS CHAPTER IS ABOUT ENDINGS that are actually beginnings, and beginnings that resemble endings. In late
March 2002, I traveled to my parents' house in Virginia to
care for my dying grandmother. My mother's mother, whom I
knew as Mémé, had advanced-stage breast cancer at the age of
eighty-one. My short stay with her was respite for my parents,
who needed to travel to upstate New York to clear out the
home of my father's mother, who had recently been moved
to an assisted living facility. I was thirty and about to return
to Montana to begin my doctoral fieldwork. When I arrived,
my grandmother had been living with my parents for several
weeks. She spent most of her days in bed in a sunny room on
the ground floor. I remember the time going by slowly. We
read, she napped, and we sat on the porch that looked over
the valley and watched spring arrive in the woods and the
fields. Sometimes she felt strong enough to get dressed and
move around, or sit at the table with a cup of tea. Often her
pain drove her back to bed, where on her bedside table a dose
of morphine waited. The morphine's oral syringe rested in
a small cylindrical ceramic vessel, white with a blue pattern

around the rim, which she had acquired as a souvenir on one of her many trips to Mexico. One day I needed to go into town for groceries, and Mémé felt well enough to accompany me. I steered her through the supermarket in the store's wheelchair and asked her what she wanted for dinner that night. "Meat loaf," she said. I didn't know how to make meat loaf and I didn't think to ask for her recipe. I made something else. My parents came home a few days later, and I flew to Montana. My grandmother said to my mother, "She was the last one." She died on April 7, 2002. When my mother called to share the news, the first thing I thought of was the meat loaf that I hadn't cooked, the simple wish unfulfilled.

When I first traveled to Orford Ness in 2012 to carry out the research I discuss in chapter 4, Grant Lohoar took us out to see the 220-year-old red-and-white-banded lighthouse at the far eastern edge of the shingle spit, north of the AWRE structures. The first documented aids to navigation on Orford Ness were two wooden towers built in the middle of the seventeenth century.[1] The doubled light allowed sailors to align the beacons to locate themselves off the coast for safe passage through the treacherous shingle and sand banks. These towers were unstable and vulnerable to undermining by processes of long-shore drift, which continually lifted the shingle and transported it farther down the spit. Although one of them was eventually replaced by a masonry structure, by the late eighteenth century the need for major improvement of the beacon system was evident. The landowner, Lord Braybrooke, commissioned the building of a new beacon set well back from the shoreline, with a lantern lit by fourteen oil lamps set in silver-plated reflectors. For a while after the new light came into use in 1792, one of the former beacons remained in place as the "low light." They were both present in the 1820s, when William Daniell featured them in an etching on his coastal circumnavigation of Great Britain and when J. M. W. Turner painted them from the vantage of a stormy sea. In 1887, the low light was decommissioned, and the high light was altered to compensate for the loss with the installation of an occulting main lantern and green-and-white sector lights.

At the time of my visit in March 2012, the 1792 lighthouse was still in operation, though Trinity House, the lighthouse authority for England and Wales, had recently determined that the light was no longer required as an aid to navigation. The background to this decision was the inexorable erosion of the bank where the lighthouse stood: Grant pointed out that the span from lighthouse base to beach crest was only fifteen meters, and with average annual erosion on the spit estimated at three and a half meters, the structure didn't have much time left. He recalled that in past years (though not recently) he had sometimes seen the base of the old low light offshore at extremely low tides; beyond it lay the ghost foundations of previous beacons. The standing lighthouse was at the eleventh hour on a slow-motion seascape clock, with the old low light already passed and the interval between each prior beacon measuring centuries rather than hours.

At that point, Trinity House and the National Trust were in active conversation about a transfer of ownership or a lease agreement, though the Trust had made it clear that if they accepted responsibility for the structure, they would take no measures to defend it. They were committed to following the guidance set out in the 2010 Shoreline Management Plan, which recommended "no active intervention" on this stretch of coastline.[2] Eventually the Grade II listed lighthouse would cede its ground to the sea. Some residents of Orford and the surrounding community had initiated a conversation with the National Trust and others about holding a wake for the structure.

Although the idea of holding a wake for a building may seem strange, the impulse to understand the decay and destruction of built structures by drawing parallels with our own corporeal vulnerability, and eventual mortality, is a very old one. Rose Macauley diagnosed the "realization of mortality" as the dominant emotion inspired by ruined and transient architectures.[3] The group of concerned local residents wanted to recognize the emotions that would be stirred up by the decommissioning and eventual loss of the lighthouse, and they began to seek funding for an extended program of events.[4] Meanwhile, although some voices within the Trust were of the opinion that the organization should take on responsibility for the structure and use it as a test case for developing imaginative and proactive coastal adaptation strategies, uncertainties about liability and expense swayed the decision. The Trust decided not to seek acquisition.[5]

Trinity House decommissioned the light in June 2013. The BBC marked the occasion with the headline, "Orfordness Lighthouse Gets Switched Off and Left to the Sea."[6] In an unexpected twist, a London lawyer with a second home near Orford purchased the structure from Trinity House in September and promptly formed the Orfordness Lighthouse Company.[7] The stated aim of Nicholas Gold's new company was "preservation so far as possible" of the beleaguered lighthouse, "until such time as it may fall victim to the sea and waves."[8] As the autumn storms began to batter the Suffolk coastline six months later, as at Mullion, it became clear that the time might be closer than anyone had imagined. Each successive sea swell and storm event brought the base of the lighthouse closer to the waves, and the newly formed company devised a hasty plan to temporarily stabilize the beach crest with a fifty-meter-long barrier of geotextile bags filled with beach pebbles.[9] The National Trust expressed concerns about the proposal, pointing out that when Trinity House announced the decision to decommission the lighthouse in 2010, they had collaboratively agreed on a position, based on the recommendations in the Shoreline Management Plan, "that we would allow natural forces to dictate the future of the building." They also noted that sea defenses might inadvertently accelerate erosion elsewhere on the spit and "cause unacceptable damage to what is a fragile habitat of international importance."[10] In a subsequent statement, the National Trust acknowledged that, given the urgency of the situation, temporary stabilization measures might be necessary to "allow time to remove the principal features and fittings from this historically important building."[11] Some of those who commented on the planning proposal seemed not to appreciate the temporary nature of the solution. One local resident opined, "All efforts should be made to prolong the life of this historic building."[12]

If we accept that our buildings have lives, then we also must accept that they, like us, have deaths. "Objects and structures that display the erosions and accretions of age seem conformable with our own transient and ever-changing selves," writes David Lowenthal.[13] Michael Shanks explores the shared life cycles of people and artifacts in a 1998 paper in which he makes "a plea for pathology."[14] "The seduction of conservation," he argues, "is one of gratification—ridding the self of this nausea of loss and decay."[15] When conservation is not an option, evidence of aging and decay is sanitized and sterilized through scientific analysis (in archaeology, these analyses include wear studies and mortuary analysis). Absent in all of this activity, Shanks avers, is the recognition that "decay is an *essential adjunct* to a *living* past . . . and a token of the *human* condition."[16] "Death and decay await us all, people and objects alike," he writes, "In common we have our materiality."[17] For Shanks, recognition of our common life cycles is a precondition for an awareness of the "symmetry of people and things," a symmetry that works to "dissolve the absolute distinction between people and the object world."[18] Our common experience of decay and morbidity fundamentally unsettles claims of objectivity and renders them relative and contingent. Shanks's insights find common ground in recent work in geography, which has also sought to question the exceptionalism of human life. We are linked through our "shared finitude," writes Pepe Romanillos, and when we extend our ethical response to nonhuman subjects, both organic and inorganic, we find ourselves in a relation of care and compassion for all vulnerable "mortal" subjects.[19]

What kind of care can we extend to a subject whose death is imminent? In clinical contexts, the term "palliative" has come to refer to care that seeks to relieve or soothe the symptoms of a disease or disorder without effecting a cure, particularly in end-of-life contexts. Palliative care of a terminally ill patient involves minimal intervention—only that necessary to ensure comfort and dignity. I once heard an interview on BBC radio with a palliative care consultant who explained that the aim of palliative care is to "help people cope with uncertainty—in the movement between life and death." During my week with my grandmother, nurses from the local hospice visited to check that she had everything that she needed, but I don't remember much detail from these visits. I remember more clearly the ritual of refilling the morphine syringe, and the daily small tasks of caring: preparing meals, washing, changing clothes. I'd like to be able to remember other things as well—reading her the poems she loved, reminiscing about her full life—but I can't honestly say that I do.

The idea of extending palliative care to buildings and to artifacts is one that was first mooted by Bob Oaks, a friend in Montana who shared with me the perplexity of finding the appropriate treatment for the derelict structures on the Montana homestead where I carried out my doctoral work. When we first took on responsibility for the site, many of the buildings were so far gone that the only sensible thing to do seemed to be to allow them to continue going and to document their gradual demise. "We can call it palliative curation," Bob suggested, which isn't as absurd as it sounds. The root of the word "curation" is the Latin word *curare*, "to tend" or "to care." The contemporary meaning, "to arrange" or "to assemble," came later. As curators of the homestead, we sought a way of respectfully and attentively easing the terminal structures into their inevitable deaths, refusing to accept the burden of guilt that would commonly be associated with the loss of a valued historic feature.

In his book *Building Lives*, Neil Harris proposes that we need "life stage rituals" for buildings as well as people. Such rituals should acknowledge "the powerful emotions raised by the expiration of a structure's time on earth"[20] and recognize that "disintegration and dissolution are part of the natural building cycle."[21] Some of the existing, implicit rituals of leave-taking include exhaustive documentation, such the early photographic surveys studied by Elizabeth Edwards, which sought to create a visual record of threatened and deteriorating architectural features. Edwards argues that the documentary impulse functions to assuage our "entropic anxieties" about disappearance and loss.[22] Such acts of recognition and revaluation can also be understood, in a more positive sense, as "life affirming" gestures for terminal structures.[23] Kevin Lynch highlights more expansive possibilities: "Since . . . the destruction and death of environment may be as significant a point in its process as its creation, why not celebrate that moment in some more significant way? . . . There could be a visible event and a suitable transformation when a place 'came of age' or was about to disappear."[24]

But any form of palliative care involves a series of decisions, and these decisions can be fraught and emotionally complex, especially if the carers are not in agreement. One of my mother's sisters felt strongly that all possible measures should be taken to extend her mother's life, and was deeply unhappy when other members of the family decided to honor Mémé's wish to stop fighting against the progress of the disease that had consumed her body. I hope I'm not taking too many liberties when I suggest that a similar conflict over end-of-life care developed on Orford Ness. Both the National Trust and the Orfordness Lighthouse Company wished to extend care and compassion to the lighthouse in its final days; they just had different ideas about how this should be done and about the appropriate rituals of retreat. The company and its supporters wished to use artificial means to provide the equivalent of architectural life support. Their temporary defenses did not promise a miracle cure, but they did seek to prolong the life of the structure "so far as possible," which may have meant months, or, more optimistically, years. The position held by National Trust, and supported by the Shoreline Management Plan, aspired to the clarity of a "do not resuscitate" order, which accepted the loss of the lighthouse as part of a natural process of erosion and landscape change. In March 2014, the district council approved the proposal, granting a permit for a maximum period of five years. The decision letter stressed the "non-permanent nature" of the engineering works and noted that future options included either "controlled demolition or dismantling" or "retention of the lighthouse structure" on another site.[25] The lighthouse would be supported with the apparatus of intensive care, but only temporarily, it seemed.

The bags were filled with pebbles and placed on the eroding edge of the bank in front of the lighthouse, but the tensions didn't fade away. The Orfordness Lighthouse Company hosted a series of public tours over the course of the summer, bringing to the Ness a steady stream of well-wishers coming to pay their last respects to the structure. The spit is usually accessible only by the National Trust ferry, but permission was granted for a limited number of additional journeys (on a vessel named *Regardless*). As the summer went on, conflicts arose between the National Trust and the company over the increased number of visitors on the site (with associated public safety risks due to the presence of unexploded ordnance in the shingle) and concerns about disturbance to the fragile, protected habitats.

With relations between the Orfordness Lighthouse Company and the National Trust at a low ebb, Nicholas Gold decided to make a formal record of his grievances in a series of members' resolutions put before the National Trust's Board of Trustees.[26] Eight resolutions appeared on the agenda for the annual general meeting in November 2014, all proposed by Gold and his supporters. The outgoing chairman of the Board of Trustees, Simon Jenkins, described the situation as "completely unprecedented"; the atmosphere in the meeting was civil but strained.[27] Several of the resolutions dealt with procedural issues: free entry, prebooking, disclosure of public funding. Others were more substantive. One resolution proposed that the National Trust should be required to consult its Architectural Panel "before any action or decision is taken by the National Trust which may materially and adversely affect any listed building within the immediate vicinity of any National Trust property."[28] Another suggested that the Trust adopt a set of "transparent and open" guidelines for local community consultation in the preparation of coastal adaptation strategies for areas of coastline threatened with erosion.[29]

In their presentation of the resolutions, Nicholas Gold and his partner, Mary Ann Gribben, referred explicitly to the disagreement over the lighthouse and its fate. They cited the conflict between "historic buildings and nature conservation" that had arisen on Orford Ness and the need for assurance that the Trust would consider the effects of their actions on structures like the listed lighthouse. Gold stated (in an apparent misrepresentation of the record) that the National Trust had been opposed to the installation of "modest sea defence," and he alleged that one of the managers had said that the lighthouse was "intended to fall into the sea, as they had a policy of, quote, controlled ruination." The company did not articulate a viable alternative plan for the long-term management of the structure, however, and in fact, at one point Nicholas Gold openly admitted, "The sea is coming in on it. . . . In a few years the lighthouse will go."[30]

During the open comments after the presentation of the resolutions, National Trust members in the audience expressed bewilderment about how local differences of opinion had escalated to the point where the proposers felt that the only course of action they had available to them was to bring the issues to the national level. One member commented, "I don't think it should ever have got this far. . . . In this case something has gone dreadfully wrong." The board was being asked to offer a second opinion on the treatment options for the beleaguered structure, but their response was equivocal: "The Board supports the spirit of the resolutions, but is unable to support the prescriptive approaches proposed and therefore recommends members to vote against them."[31] The Orfordness Lighthouse Company's proposal to proscribe the National Trust's involvement in the patient's care plan was deemed inappropriate, and the resolutions were rejected by a narrow margin.

Reading between the lines, I wonder if some of the apprehension directed toward coastal policy at Orford Ness is related to the repeated reference to the forces of "nature." The stated desire to allow "natural forces to dictate the future of the building" is perhaps perceived by some as a way of sidestepping the fraught issue of who will decide whether, and how, the lighthouse stands or falls. Nature is apparently framed as the primary agent of change and destruction, with people standing aside to let it have its way. Of course, the representatives of the Environment Agency (responsible for shoreline management planning) and the Trust would be the first to admit that there is no real possibility of standing back and allowing for the uncontrolled destruction of the structure as the sea undermines the ground it stands on—not least because it is possible that the collapse of the lighthouse tower in situ could create an artificial groin, with the potential to alter coastal process along the whole of the spit.[32] On the other hand, despite the Trust's attempts to open up the conversation, no one has articulated the practical steps that will need to be taken to allow the lighthouse to make the transition from *here* to *gone*, and communication about how the erosive process initiated by the sea will be completed by deliberate acts of human intervention and remediation remains constrained by underlying tensions and differences of opinion.[33]

These issues are caught up with deeper considerations around the underlying character of the forces that are now accelerating coastal processes and sea level rise around the planet. The Anthropocene epiphany reminds us that we are deeply implicated in earth processes all the way down—in soil layers, tree rings, rising tides, and swelling storm surges.[34] It follows, perhaps, that we have a responsibility to the environments and entities that have been transformed by our actions. In this light, a focus on palliative care for threatened structures forces us to accept this responsibility and to seek new ways of conceiving of care in relation to transient and transitional places and things. Joan Iverson Nassauer and Julie Raskin, in their work on the ecologies of abandonment in Detroit, have written about the social and psychological benefits generated by small acts of ordering and maintenance in otherwise derelict landscapes.[35] Care, in this mode, is not so much extended to (animate and inanimate) others but produced with them.[36]

Other cultural contexts suggest alternative templates for how retention and relinquishment can be brought into productive relation. The *malanggan* tradition discussed in chapter 2, which performs remembrance through material transience, is one model. Parallels can be found in East Asian architectural practices, which tend to privilege the transmission of spiritual significance across generations rather than the material permanence of a built structure.[37] The organic decay of structural material, and its cyclical repair and replacement, is given meaning through grounding in traditions that embrace impermanence, renewal, and rebirth.[38] The elusive Japanese concept of *wabi-sabi*, described as "an aesthetic sensibility that finds a melancholic beauty in the impermanence of all things," values transience as a reflection of the irreversible flow of life and matter.[39] The anxiety about impermanence that characterizes modern Western heritage practice is alien to many other cultural traditions.

David Lowenthal cites a comment offered by the inhabitant of a damaged house in Santa Clara Pueblo: "It has been a good house; it had been taken care of, blessed and healed many times in its life, and now it is time for it to go back to the earth."[40] While we cannot, as Lowenthal warns, simply shed our obsession with material preservation to try on borrowed cultural practices, awareness of "different modes of defining and preserving pasts . . . may help us to extend the forms and functions" of our own.[41]

The Ise Shrine in Japan has been rebuilt every twenty years for many centuries. The twenty-year interval aligns with that of a human generation, but it also corresponds to the life cycle of the deities that inhabit the structure and the onset of decay in the temple's supporting columns.[42] An analogue understanding could frame the Orfordness Lighthouse as a structure that is also, similarly, continually renewed, though at 200-year, rather than twenty-year, intervals, to correspond with rates of coastal erosion. In the most recent iteration of the cycle, the chain of material beacons at this site is broken, replaced by the ethereal (though no less material at its source) technological warning system of GPS and the strengthened beam at the nearby Southwold light. The emotional response that the imminent demise of the lighthouse has generated suggests that we may be more inclined that we would admit to think about the objects we share our lives with as living entities.[43] Paradoxically, however, our impulse to seek material preservation can work to disrupt rather than extend the life cycle, by seeking stasis and effectively embalming a living thing. Sven Ouzman writes of the "beauty in letting go" of some objects when it is culturally appropriate to do so.[44]

In Euro-American heritage contexts, practices that allow us to mark material transformation intentionally and attentively are only beginning to emerge. This may be because one of the defining qualities of palliative care is its ambiguous relation to finitude and absence. We can only fully realize the significance of the care extended after the threshold of death has been passed. The ritual of caring by definition must contain with it an element of denial, hope, or both. Because we cannot adequately countenance death, we cope with anxiety through intervention—attempts to control the uncontrollable, to predict the unpredictable. As we struggle to find the most sensitive way to attend to and communicate with dying others, we always come up against the "the unsettling force of absence inscribed within the conditions of communication through which we are able to negotiate death and dying at all."[45] An "anticipatory mourning" haunts and troubles the very conditions of communication.[46]

The negotiation of the process of dying—whether of a person or of a place—is also almost always bound up with the exercise of power, whether soft or hard. Who decides when death will be deferred, when it will be resisted, and when resuscitation will be attempted? The morbidity of places can be catalyzed by many different forces—the withdrawal of capital, human migrations, acts of war, natural disasters, industrial accidents.[47] While the situation with the lighthouse is relatively clear-cut, how do we apply these ideas to sick places, like Fukushima or Chernobyl, or the hollowed-out cities of the postindustrial north of England? The range of available responses to the recognition of dying places is limited. In most contexts, abandonment is framed as a problem requiring a solution in the form of remediation, redevelopment, or regeneration, or conservation and preservation. Meanwhile, inevitably, new practices emerge in the place of the extinguished ones, and life goes on (though often in illegible or illegitimate forms). There is little willingness to accept abandonment as a valid phase in the life cycle of a place—a reluctance that also arises in attitudes toward human health.

Paul Harrison writes, "As if mirroring the reductive and clinical nature of biomedical science, corporeal vulnerability more often than not appears as a problem to be solved rather than as an inherent—and inherently significant—condition of existence."[48] The inertia of intervention privileges the "ongoing process of holding together" over the imperatives of entropic process.[49] But as I've been arguing throughout this book, in some places, in some times, it is possible to imagine the contours of another mode of attention that involves care without the attempt to control and that proposes that apparent loss can also be generative of something new. Paul Kingsnorth describes this as a process of "falling away": "Lose something, let go of it as it falls away, and you may gain something else. Or you may not, but at least if you have let go, said your goodbyes, accepted your given load—then maybe you can watch it fall with lighter shoulders."[50]

Over the past several years, rituals of leave-taking for the lighthouse, many of them supported by the National Trust, have begun to demonstrate what this new mode of attention might look like as well as the role that art and creative practice can play in helping people assimilate change.[51] Simon Read, a painter who lives on a barge not far from Orford Ness, used the predictions of coastal erosion and sea level rise in the 2010 Suffolk Shoreline Management Plan to depict the lighthouse as a red dot bobbing in the surf off the new coastline, sending out a ghosted halo of lost light. Thomas Dolby, a nearby resident, produced a film entitled *The Invisible Lighthouse*, which filters the lighthouse's future through a dark rendering of destructive energies held within the wider landscape of the Ness. Liz Ferretti, director of the Orfordness

Lighthouse Project (which developed out of the early conversations about the need for a lighthouse wake), recently hosted
a short story competition that encouraged people to address
the uncertain future of the symbolic structure. In autumn
2015, she worked with local schoolchildren and other artists
to create a "cycle of lighthouse songs," inspired by the structure's history and looking to its future, which was performed
in Orford Church (in association with an exhibition of recovered artifacts and related artwork).[52] The intention of these
activities, Liz explains, is to allow people to examine their
emotional response to the loss of a loved local landmark, but
to do so in an imaginative and oblique way that sidesteps the
fraught negotiations over its immediate fate.[53]

After the loss comes the grieving, though anticipatory mourning often begins long before the end has been reached. A
range of emotional responses characterize our experience of
grief. Elisabeth Kübler-Ross named these emotions—denial,
anger, bargaining, depression, and acceptance—but she was
careful to point out that the progression through these stages
is not necessarily sequential.[54] Richard Hobbs has diagnosed
expression of these emotions in the responses of ecologists
to species extinction and landscape change. When people are
caught in different phases of the grieving process, he notes,
conflict can arise. Those in the bargaining mode try to hedge
against future losses through deliberation over trade-offs
and priority setting (or adoption of short-term measures to
provide temporary stability).[55] Others, who have accepted the
inevitability of loss and change, may be more willing to work
on finding ways forward in the new circumstances.[56] Kathyrn
Yusoff, also referring to species extinctions, observes, "Loss
requires mourning and grieving for the destruction of a relation and those subjects that are constituted through that

relation."[57] Yet on the other side of loss, there is always the possibility that new relations, and new subjects, will emerge.

A few weeks after Mémé's death, the family held a memorial gathering in Virginia. My grandmother's six children and six grandchildren were all present, and my sister and sister-in-law carried two grandchildren-in-waiting, who would be born that summer. We gathered in the back garden below the porch where my grandmother and I had sat together a few weeks earlier. The ceremony was brief and simple, centered around the planting of a young maple tree. We tipped some of Mémé's ashes into the hole in the earth before we planted the tree, and my mother claimed later that she saw the glint of Mémé's gold wedding ring as it fell. All of my emotions had burrowed underground that day, gone somewhere else. I remember thinking abstractly about the contrast between swelling life (the exuberant Virginia spring, my sisters' pregnancies) and stark loss. I remember thinking how sad it was that my grandmother would never meet those unborn children and feeling a stab of insufficiency when I realized that I didn't have a child to bring to the gathering. It wasn't until I started writing this chapter that I was reminded of the date of the memorial, April 20, 2002. My son was born exactly five years later, on April 20, 2007. Maybe he was there too, the inverse of an echo, the presence that casts itself back from the future. My son now uses the Mexican morphine cylinder to hold his paintbrushes. It's been broken several times and patched back together, and carried on at least three cross-Atlantic migrations. If people make things and things make people, then that small vessel must hold some of my grandmother—and now some of us.

There is a scene in Peter Greenaway's 1985 film *A Zed and Two Noughts* that takes place in the lab facility of an unnamed zoo. Two brothers, Oliver and Oswald, both zoologists employed at the zoo, have recently lost their wives in a car accident that involved a fatal collision with a swan. After the accident, both brothers become obsessed with decay, and Oliver sets up a series of time-lapse cameras in the zoo labs to document the gradual decomposition of various entities: first an apple and a bowl of prawns, and later a juvenile crocodile, the culpable swan, a road-killed dog, and a zebra. In the scene, the room's darkness is punctuated by irregular flashes of light that illuminate the subjects in polythene enclosures. The progress of putrification and the breakdown of each body is measured against a gridded background. Oliver comments, "I sit here for hours. It's like sitting amongst lighthouses. Each lighthouse is giving you a bearing on lost spaces of time. There are tens of thousands of photographs taken here, all taken very patiently. Because decay can be very slow. Nine months for the human body, they say."[58]

We need to "think of a world not of finished entities . . . but of processes that are continually carrying on," asserts Tim Ingold, and we need to "think of the life of the person, too, as a process without beginning or end, punctuated but not originated or terminated by key events such as birth and death, and the other things that happen in between."[59] How can we understand the lighthouse as a process rather than a thing? Ingold points out that our convention is to pinpoint age to the moment of the making of an object or structure, as with the birth of an individual person. But what if we allow the life cycle of the lighthouse to scroll out into the prehistory of its construction in 1792, and seek to tell the life stories of its constituent materials as well? The iron, brick, and concrete that make up the bulk of the structure have biographies of formation, extraction, and transformation that precede their assembly in the ostensibly coherent shape of the lighthouse, and these constituent materials also have a future, which will play out after the structure loses its current form. We are willing to accept that the integrity of the shingle spit endures, despite its continual reshaping and reassembly, pebble by shifting pebble. Why not extend this sense of dynamic, distributed identity to the lighthouse as well, and, as Ingold suggests, locate meaning in "persistence, not preservation"?[60] Such a shift in perspective would allow us to focus not on the material persistence of the heritage object in its original structure and form, as explored by Tim Cresswell and Gareth Hoskins, but on the persistence of the matter incorporated within it, as it is drawn into other systems and processes.[61]

But how do we navigate the transition? The rituals of remembrance proposed for the lighthouse have been reflective and reserved, at a remove from the physicality of the structure itself. The end for the lighthouse, by necessity, is likely to be aggressive and abrupt; it will involve radical material intervention in the form of heavy machinery and wrecking tools. It will need to be dismantled through an act of deliberate destruction that may involve some attempt at salvage, but it is unlikely to be gentle, or even very careful.[62] To return to Yusoff, the unmaking of the lighthouse will activate an "aesthetics of loss" that is premised on violence and will catalyze "a different ontological configuration than care . . . beyond affirmative relations."[63] Jens Wienberg has explored a similar dilemma that is playing out in relation to threatened coastal structures in Northern Jutland, Denmark. He proposes the concept of "creative dismantling" as a compromise between preservation and destruction, which seeks to generate new knowledge through acts of salvage, displacement, and reuse.[64] As Tim Flohr Sørensen points how, however, this model still stops short of realizing the radical potential for a sacrificial heritage logic, which sees the process of decay and disappearance as having value in its own right.[65] Models for how the demolition process might be mediated and made meaningful are perhaps best sought not in conservation practice but in the brutalist architectural interventions of artists like Robert Smithson and Gordon Matta-Clark.

Smithson, who approached "entropy as the repressed condition of architecture," conceived of several works that attempted to express this principle, some of them never realized.[66] In 1970, he put forward a proposal for an *Island of Broken Concrete* (also called *Island of the Dismantled Building*) on a barren outcrop in Vancouver Bay. This piece, which would have involved covering the rock with a layer of concrete rubble, was meant to symbolize intentional obsolescence and the deliberate production of ruination.[67] In the same year, he explored the process of dearchitecturization with the creation of *Partially Buried Woodshed* on the Kent State campus. In these works, as Yve-Alain Bois and Rosalind Krauss write, "architecture is the material, and entropy is the instrument."[68] Smithson's ideas about architecture and entropy were taken up and extended by Matta-Clark, who developed his own radical practice by deconstructing and dismembering structures that were slated for demolition. *Splitting* (1974) sliced a suburban house in half, and in doing so exposed the fundamental ephemerality of architecture by making visible its contingency and impermanence.[69]

Would it be possible to, in Smithson's terms, "accept the entropic situation" that defines the future of the Orfordness Lighthouse?[70] When the lighthouse was originally built, the rotation of the light was achieved by running a clockwork mechanism, which was manually wound by the lighthouse keepers. The complex clockwork was housed in a cast iron column that rose the height of the tower. The shaft is still in place, and a brass plate offers detailed operation instructions: "To start the apparatus: lower the clutch, release the clock brake, gradually rotate the apparatus with the hand-turning gear. To stop the apparatus: raise the clutch, apply the clock brake, let the apparatus run free until it comes to rest."[71] One proposal for radical persistence, rather than rigid preservation, might involve the salvage of the iron shaft and its reinstallation inland (perhaps on one of the disused concrete platforms that date to the MoD occupation of the site). The original clockwork could be restored to animate a new beacon feature (using the salvaged sector light?) on the top of the shaft, a memory machine that runs on the physical exertion of individual human bodies.

Although traces of the lighthouse's material presence will no doubt persist in more or less legible forms, the lighthouse itself will pass into oblivion. Other inhabitants of the disintegrating Suffolk coast, not far from Orford Ness, are well versed in the paradox that arises when absence, not presence, underlies shared structures of feeling. In the twelfth century, Dunwich was a significant political and ecclesiastical center; now, most of the former town lies under the sea. Benjamin Morris has written about how, in this place, identity is defined by erasure, and the remnant community grapples with a sense of continuity premised on mortality.[72] Places like Dunwich, and like the ground soon to be lost under the lighthouse, force us to give witness to absence and negativity, against a powerful cultural narrative that stresses the opposite.[73] Morris, citing Walter Benjamin, asks us to find "a new beauty in what is vanishing."[74]

8

Beyond Saving

CARE WITHOUT CONSERVATION

Remember my little granite pail?
The handle of it was blue.
Think what's got away in my life—
Was enough to carry me thru.

Lorine Niedecker, "Remember My Little Granite Pail?"

WHEN I WAS CARRYING OUT my doctoral fieldwork, I spent an afternoon with the curator at the Grant-Kohrs Ranch, a property in Montana now managed by the National Park Service. During my visit, the curator showed me some of her field collections. The items in her care included tin cans full of nails, mismatched horseshoes, unidentifiable metal parts, stray hardware, lengths of rope, and other mundane objects associated with late nineteenth- and twentieth-century farming and ranching practice. I asked the curator how the public valued these collections, given that the material they contained would seem unremarkable to anyone who had spent time in the rural American West. Her answer was instructive:

> The thing that seems to make people respect it more is . . .
> the better it's taken care of. If it's all in a heap they consider
> it trash. But if you single it out and put it in a little tray and

> pad it and look like you're trying to take care of it, it seems
> to have more value, in people's eyes.[1]

The curator's comment aligns neatly with an observation made by Cornelius Holtorf and Oscar Ortman: "We prefer ... a past that is fragile, cannot be replaced, and needs our help. ... One might even say that archaeological sites are not being saved because they are valued, but rather they are valued because they are being saved."[2] The preservation paradigm that guides most contemporary heritage practice singles out certain features, puts them in various equivalents of the "little padded tray," and makes it look like we're trying to take care of them.[3] In this sense, the act of extending care actually produces value, although it is often presented as a response to the inherent value of the threatened object or structure.

One French architectural historian describes the impulse to rescue threatened features as part of a "Noah complex," which frames the material past as always endangered, requiring intervention to avert loss (and providing a circular justification for further investment in preservation measures).[4] Holtorf returned to these themes in a recent essay, written with Anders Höberg, in which they argue, "As a result of preoccupation with our all-too-human needs and desires to care, and to give the impression that we care, we have never asked what role we can expect heritage to play in the *actual* future."[5] In fact, our efforts to preserve as much as possible might backfire, given that future generations may perceive as less valuable what is less rare, and an abundance of preserved heritage sites and features may inspire indifference rather than the intended appreciation.[6]

As the preceding chapters have shown, however, the impulse to care is not so easily extinguished. Even when a decision has been made to accept eventual ruination, as at

Mullion Harbour, in moments of threat, it is extremely difficult to step back and allow destruction to continue unchecked. Letting "nature take its course" is invariably more
workable in theory than in practice. If we are to explore alternatives to the preservation paradigm, perhaps we need to
develop modes of care that help us negotiate the transition
between presence and absence. Greg Kennedy, drawing on
both Heidegger and the teachings of Buddhism, offers insights that might be useful in this regard. He makes a distinction between care that imposes its will on an external world
of things and beings, and care that establishes a relation with
the cared for, and allows that relationship to work back on the
self in unpredictable ways. He writes, "Things are disclosed
as things only by our taking care of them in a manner that allows them to refer their being back to our essential embodied
neediness."[7] His use of the term "neediness" corresponds to
an acknowledgment of human finitude and vulnerability, and
the willingness to recognize the same qualities in nonhuman
subjects. He states, "Authentic care senses the truth of death
and discloses it accordingly."[8]

Although Kennedy's argument is directed toward rethinking our relationship with the disposable, it offers useful resources for grounding an entropic heritage practice, in which
the withholding of physical care does not have to mean withdrawal of a care-ful attitude toward the objects of the past
that we engage with. The key, it seems, is to realize that by accepting ongoing process, we are not automatically triggering
disposal and loss. Rather, we may in fact be opening ourselves
up to a more meaningful and reciprocal relationship with the
material past. Kennedy writes, "Taking care of a thing in a
way that lets it be what it is acknowledges, even if only tacitly,
that the thing shares the same essential fragility of our embodied existence . . . What practical taking care acknowledges

is the tendency of all physical beings to degrade, decay, to lapse into nothingness."[9] Sometimes practical taking care may involve acts of repair and maintenance that secure the material fabric of the thing; at other times, taking care may involve withholding repair and letting the thing carry on with its changes. If we choose the second path, we may find that it offers the opportunity to recognize the interpenetration between ourselves and a wider world of beings in a "network of mutual relations."[10]

Castle Drogo looms over the Teign Gorge at the edge of Dartmoor, a vast granite folly often described as the last castle in England. Millionaire Julius Drewe built the castle in the first decades of the twentieth century as an invented ancestral seat to house the fortune he had amassed through his prosperous retail empire. He hired architect Edwin Lutyens to design the structure, which incorporated approximately 5,000 tons of local granite and used imported Trinidadian asphalt on the flat roof. The roof leaked, and so did the windows. The National Trust acquired the house and its extensive grounds in 1974; after three and a half decades, the building's porosity had become so problematic that they decided to launch a last-chance campaign to raise the funds necessary to "Save Castle Drogo" from "certain ruin."[11] Eleven million pounds later, in 2012, construction commenced. The whole structure was enclosed in scaffolding and a vast white plastic tent. The castle remained open to visitors, although most of the collection was boxed up and many parts of the castle were off-limits. During construction, the National Trust commissioned artists to make work in and around the site, in response to both the history of the building and to its contemporary remaking. At the midway point of the five-year restoration project, a group of artists was selected to imagine what might have

happened if Castle Drogo had not been saved and had instead been "left to the elements."[12] The resulting work offers
a glimpse of what it might look like to invite the "network of
mutual relations" into the space of heritage interpretation.

Off a first-floor corridor, just past an installation com-
memorating *The First Drip* to penetrate the building, is
The Outside In Room.[13] The former night nursery has been
transformed to blur the boundary between inside and outside, and to introduce visitors to the microinhabitants that
conservation usually seeks to eradicate. Moth-eaten curtains
hang off the four-poster canopy bed, and the flooring has
been treated to simulate the effects of woodworm. A false
wooden wall covers one side of the room, its weathered slats
twined with ivy and ferns and its surface perforated to let
through simulated natural light (and a glimpse of the Dartmoor countryside on a digital screen). Throughout the room,
giant fabric models of common domestic insect pests have
been placed on tables and mantles—wood lice, silverfish,
case-bearing moths, furniture beetles—alongside short de-
scriptions of their diets and life cycles. A framed fragment of
moth-eaten carpet hangs above the fireplace.

A Little Cupboard of Decay provides a filmic narrative to
help the visitor make sense of all this. The film, shown on a
screen set into a wooden cabinet, draws a parallel between the
erosive processes that shaped the Dartmoor landscape and
the unwelcome, moisture-induced erosion of the castle itself
before going on to introduce the "microbes and mycelium
spores" and the ranks of "tiny decomposers" working away at
the fabric of the structure. The narrative points out that the
insect names provide evidence of the long-standing relationship between people and pests: clothes moths, flour beetles,
fur beetles, bed mites, grain weevils, wine moths, book lice.
Each agent of decay is celebrated as a "highly evolved expert

in its individual field," though it is noted that their expertise is appreciated more outside (where they provide essential services to "the rich cycles of Life on Earth") than in (where they are "the Heritage Industry's greatest foes"). *The Outside In Room,* the film explains, allows them to enjoy "an interrupted paradise of decay" until the "heroic saviours" working on the restoration of the building summarily evict them.

The provisional and playful installation makes space for the other beings that inhabit the castle within the context of the restoration that will ensure (if only temporarily) their expulsion. As the conservation project nears completion, it becomes possible to contemplate a counterfactual trajectory of nonintervention. Interpretation functions in the tense of the speculative future anterior, articulating "what would have been," had the roof not been replaced, the building not "saved." It is possible to read *The Outside In Room* as a vindication of the necessity of repair (which is partly what the National Trust intended), but it is also possible to read it against the grain as a celebration of another set of choices. When the fund-raising campaign launched, some suggested that the building be left to ruin, given its relatively recent pedigree and the pressing need for funds to save other (presumably more worthy) structures. The prospect of the castle's demise is written into its history, from its first leak, and by acknowledging this reality, the recent interpretation "opens up a time and space for the monument that does not relegate it to a past that is already accomplished, nor to an anticipated future."[14] The artwork instead, as Aron Vinegar and Jorge Otero-Pailos have written, "engages [by] mobilizing the possibilities inherent in the rhythms, echoes, resonances and staging of [the building's] complexity."[15]

The work at Castle Drogo, in a tentative way, shows a willingness to engage with the potentially generative aspects of entropy and decay that is unusual in contemporary heritage practice. In terms set by Elizabeth Grosz, it works in the "in-between," finding meaning in the collapse of boundaries that are usually taken for granted. Grosz writes,

> The space of the in-between is the locus for social, cultural and natural transformations: it is not simply a conve-nient space for movements and realignments but in fact is the only place—the place around identities, between identities—where becoming, openness to futurity, outstrips the conventional impetus to retain cohesion and unity.[16]

So we return to the question of identity and subjectivity introduced in the first chapter and touched on since in various ways. To open ourselves up to a space of "movements and realignments" is to unsettle our own sense of a coherent and unified self, to recognize that our identities are made through processes of subversion and fraying as much as they are through processes of consolidation and stabilization. When we accept the continual becoming of the objects and architectures we share our world with, beyond a narrow conception of their instrumental value, we also acknowledge our own becoming. In the process, our sense of temporality shifts to allow the past to fold into the present in indeterminate ways. Michael E. Zimmerman, writing about Heidegger's concept of the self as "the clearing in which entities appear," comments, "Understanding occurs [not as a relation between mind and object, but] because human temporality is receptive to particular ways in which things can present or manifest themselves."[17] By decentering and dissolving the mind–object relation, he argues, we are able to free ourselves for "spontaneous compassion towards other beings, human and nonhuman alike. One 'lets things be' not for any external goal, but instead simply from a profound identification with all things."[18]

The act of "letting be," when performed intentionally and attentively, can perhaps form the foundation for a post-humanist heritage paradigm. Rodney Harrison touches on the ethical implications of such a shift in perspective and practice, arguing for "a more inclusive sense of ethics, that acknowledges not only the rights of humans, but also those of other-than-humans—agentive animals, plants, objects, places."[19] Such an ethical stance, he suggests, may free us from the compulsion to instinctively conserve and instead allow us to find, or create, modes of action appropriate to specific and unique circumstances. He draws on Deborah Bird Rose's concept of "connectivity ethics," which she defines as "open, uncertain, attentive, participatory, contingent."[20] When encountering a vulnerable other (a building, a species, an artifact, a place), one is called on to act, but the appropriate action is allowed to emerge from the encounter, and it may be that our sense of responsibility leads us to attend to change and transformation rather than revert to perpetuation and preservation. We can begin to imagine what it would feel like to extend care without conservation—and to unshackle ourselves from the instinctive leap to save at all costs.

Attending to processes of decay and disintegration can be as productive of heritage values as acts of saving and securing, but these may be different values than we are used to identifying with heritage practice. Ioannis Poulios has called for a new conception of "living heritage" as an alternative to dominant heritage models, which privilege the preservation of original fabric and function by establishing a discontinuity between the past and the present.[21] Living heritage instead privileges "change, in the context of continuity" and makes space for a much broader range of material practices, which might include replacement, renewal, and attrition.[22] Such an approach rejects the premise that heritage is by definition a nonrenewable resource and instead asserts that heritage can be continually renewed if the social relations and practices that give it meaning are sustained over time, even if the associated material fabric is substantially altered or erased. The process of transformation can be productive in its own right: some things will remain, but others will be allowed to pass on, or over.

To imagine how a postpreservation heritage practice might unfold in a specific place and make possible particular futures, let us return to Mullion Harbour. Before moving forward, however, we need to take a snapshot from the recent past. On March 13, 2015, two entries appeared on the Lizard National Trust Facebook page (Lizard referring to the name of the peninsula where Mullion is located). The first entry included four photographs of the Mullion repairs, captioned "definitely on the homeward straight now": one documented the repaired southern breakwater and its newly smooth concrete flanks. The other entry reported on progress repairing Tremayne Quay, another National Trust property located several miles away in the sheltered upper reaches of the Helford River. This entry noted, "Thanks to some surplus coping stones from Mullion Harbour . . . we have been able to rebuild a section of wall in the same style as the rest of the quay." The long, rectangular granite blocks had been displaced by the use of concrete in the Mullion repairs, making them available for other uses. It is not certain whether the coping stones were part of the original fabric of Mullion's structure, given that they were located on a section of breakwater that has been rebuilt multiple times in the past. Storm-sheared sections of Mullion's metal railings were also repurposed into new mooring posts for Tremayne Quay.[23]

As Mullion is gradually dismantled over the next few decades—or as its fabric undergoes an incremental replacement of granite with concrete, akin to the renewal of cells in a body—its materials will gradually be released, and often reused. The movement of the stone, in particular, presents an opportunity to narrate both the history of the harbor, when the stones were assembled into the structure, and its future trajectories, when the same material will be assembled into other (equally ephemeral, but provisionally durable) structures. The traveling stones could be interpreted in situ, in their new homes, with a plaque noting the path of their travels. The memory associated with Mullion Harbour would become mobile, expressed through a material link, but not reliant on that material presence for its persistence. As Kevin Lynch reminds us in his brilliant 1972 book *What Time Is This Place?*, "Preservation is not simply the saving of old things but the maintaining of a response to those things. This response can be transmitted, lost or modified. It may survive the . . . thing itself."[24] Back in Mullion Cove, as the smaller granite setts are once again lifted from the harbor walkways in subsequent storm seasons, and as a decision is ultimately taken not to replace them, they might be gathered and reassembled elsewhere in chance cairns, perhaps on the hillside overlooking the harbor, where over a dozen memorial benches already cluster. The cairns would join the benches in performing the work of memory, not through the promise of presence, but through, as John Wylie has observed, a constitutive absence.[25]

A heritage practice that places process on an equal footing with preservation would need to cultivate a greater willingness to work with fragments and would need to seek alternatives to reconstruction and restoration. To return to Riegl's terms, interpretation would need to be attentive to the transition from extensive effect (the perception of the whole) to intensive effect (the force of the fragment).[26] It could dwell in the interval in which fragmentariness is actually an interpretive asset rather than a perceived deficit. As Mats Burström writes, it is "the lack of a complete original that fascinates people and invites interpretation . . . [and] gives people an active role in the interpretive process."[27] Our perception of the part, rather than the whole, opens up a space that invites speculation and connection. Burström goes on to claim, "Things may gain rather than lose meaning through fragmentation, and for this reason fragmentation may be intentional rather than accidental."[28]

Intentional fragmentation is one in a range of alternative practices that could emerge in the future in relation to certain sites and subjects. It is perhaps unlikely that a shift toward curated decay will displace the preservation paradigm anytime soon, but there may be opportunities to use the ideas in this book to frame experiments that work with abandonment and to stage ephemeral interventions that respond to moments of flux or change. Jane M. Jacobs and Stephen Cairns have written of the creative possibilities that coalesce around informal and incremental architectures, in which people are granted the agency to respond to change and ruination on their own terms, unscripted.[29] At the moment, our comportment toward heritage objects tends to cleave to a relatively narrow register of possible responses—appreciation, contemplation, concern. A postpreservation model of heritage would open up many more, and many of them in an active rather than a passive mode of engagement—creation, cultivation, improvisation, renewal.

As I bring this book to a close, I hope I have stimulated some curiosity about what it might look like to test some of the ideas I have introduced in a broader arena. But I can't end the story without acknowledging the unanswered questions that have been stirred up and are now swirling in the murk. How would heritage legislation and policy need to change to accommodate these approaches? What are the political implications of providing (or appearing to provide) a justification for neglect and disinvestment? Could institutional heritage practice adopt forms of care that make no claims to material protection, or is the risk of loss (of both reputation and resources) too great? Can designation countenance destruction? Most of these questions are outside the scope of this book, although I'm continuing to work through them and hope to be able to offer some tentative answers before Mullion Harbour disappears (though perhaps not before the Orfordness Lighthouse does). The aim of this book has been to offer glimpses of how it could be otherwise, and to draw out how what we might call entropic heritage practice is already emerging in certain places and circumstances, although it may not be known as such. I'm not able to follow these places through to their next chapters, so I need to leave them in the midst of their changes, with a final invitation to think about what could be gained if we were to care for the past without pickling it.

Acknowledgments

Curated Decay has materialized over many years, in conversation and collaboration with numerous people. Some people have been part of the story from the beginning; others made a more discrete contribution, with a passing comment on a walk or an identification of an obscure plant. I'm grateful to all of you for sharing the places and the puzzles that became this book: Jon Bennie, Steve Bond, Nigel Clark, Tim Cresswell, Bill Cronon, Tim Dee, Sarah DeSilvey, Erin Despard, Tim Edensor, Bob Felce, Liz Ferretti, Michael Gallagher, Bradley Garrett, Pete Hallward, Volker Hecht, Dan Hicks, Sandra Koelle, Hayden Lorimer, Antony Lyons, Elizabeth Masterton, Misha Myers, Bob Oaks, David Papadopoulos, Simon Read, Gillian Rose, Kerri Rosenstein, Matt Thompson, Christian Weis, Louise K. Wilson, and Kathryn Yusoff. I'm also indebted to the generosity and honesty of the people who took the time to share their insights into the work they do (and gently corrected my interpretation of the same, when necessary): David Bullock, Alastair Cameron, Maria Craig, Phil Dyke, Chris Ford, Rebecca Glover, Dave Hazlehurst, Duncan Kent, Grant Lohoar, Nick Marriott, and Justin Whitehouse.

Over the past eight years, I've been a fortunate participant in a number of research communities; each provided essential context and constructive critique as I carried out the investigations that have found their way into this book. Fellow members of the Geographies of Creativity and Knowledge research group at the University of Exeter, including my stellar

Ph.D. students, offered support and sustenance over the long haul. The Ruin Memories team, convened by Bjørnar Olsen (with support from the Norwegian Research Council), provided just the right mix of sympathy and skepticism when I needed it most. More recently, my collaborators on the Heritage Futures team, led by Rodney Harrison (with support from the Arts and Humanities Research Council), encouraged me to begin to think about how these ideas might travel beyond the boundaries of the sites described in this book. I'm also appreciative of the people who listened to me talk about this research over the years in Aberystwyth, Bayreuth, Chicago, Edinburgh, Exeter, Falmouth, Glasgow, Leicester, London, Los Angeles, Madison, Milton Keynes, Missoula, Reykjavik, and Santiago de Compostela.

A few wise and generous souls took the time to read early drafts of the entire book and offered extremely helpful advice on revision and reframing: Nadia Bartolini, Jeremy Clitherow, Cornelius Holtorf (who read three chapters while sitting in the ruins of the Colosseum), Gareth Hoskins, and James Ryan each should take credit for nudging the final draft into a more sensible form. Jason Weidemann and Erin Warholm-Wohlenhaus with the University of Minnesota Press skillfully steered the book through its validations and transformations. Finally, boundless thanks to Russ, for keeping the home fires burning, and to Ronan and Leif, for living the changes with us.

Notes

1. Postpreservation

1. In 2015, a newly formed body, Historic England, inherited English Heritage's position as the U.K. government's statutory adviser and a statutory consultee on all aspects of the historic environment and its heritage assets.
2. Rodney Harrison, *Heritage: Critical Approaches* (London: Routledge, 2012), 7.
3. Cornelius Holtorf, "The Heritage of Heritage," *Heritage and Society* 5, no. 2 (2012): 153–73.
4. Tim Winter, "Clarifying the Critical in Critical Heritage Studies," *International Journal of Heritage Studies* 19, no. 6 (2013): 532–45.
5. Graham Fairclough, "Conservation and the British," in *Defining Moments: Dramatic Archaeologies*, ed. John Schofield (Oxford: Archaeopress, 2005), 158.
6. Legislation included the United Kingdom Ancient Monuments Consolidation and Amendment Act of 1913 and the United States Antiquities Act of 1906.
7. Maria Balshaw, "The (Heritage) Elephant in the Room," Heritage Exchange 2014, http://www.heritageexchange.co.uk/.
8. Rodney Harrison, "Forgetting to Remember, Remembering to Forget: Late Modern Heritage Practices, Sustainability and the 'Crisis' of Accumulation of the Past," *International Journal of Heritage Studies* 19, no. 6 (2013): 579–95. See also Michael Landzelius, "Commemorative Dis(re)remembering: Erasing Heritage, Spatializing Disinheritance," *Environment and Planning D: Society and Space* 21 (2003): 195–221.
9. Mark Augé, *Oblivion,* trans. Marjolijn de Jager (Minneapolis: University of Minnesota Press, 2004), 89.
10. Adrian Forty and Suzanne Kuchler, eds., *The Art of Forgetting* (Oxford: Berg, 1999).
11. Þóra Pétursdóttir, "Things Out-of-Hand: The Aesthetics of

Abandonment," in *Ruin Memories: Materialities, Aesthetics and the Archaeology of the Recent Past*, ed. Bjørnar Olsen and Þóra Pétursdóttir (London: Routledge, 2014), 338.

12. Cornelius Holtorf, "Averting Loss Aversion in Cultural Heritage," *International Journal of Heritage Studies* 21, no. 4 (2015): 405–21.

13. Many of the sites that I discuss in this book are managed by the National Trust, a charitable U.K. body founded in 1895 and now responsible for looking after over 600,000 acres of land; 900 historic houses, gardens, parks, and former industrial sites; 149 museums; and 775 acres of coastline in England, Wales, and Northern Ireland. Although I worked closely with many National Trust employees in the course of writing this book, the project has not been endorsed by the organization in any way.

14. Dougald Hine, "Remember the Future?," *Dark Mountain* 2 (2011): 264.

15. Jane Bennett, "The Force of Things: Steps towards an Ecology of Matter," *Political Theory* 32, no. 3 (2004): 350.

16. Ibid., 351.

17. Luke Introna, "Ethics and Flesh: Being Touched by the Otherness of Things," in Olsen and Pétursdóttir, *Ruin Memories*, 51.

18. Nick Bingham and Steve Hinchliffe, "Reconstituting Natures: Articulating Other Modes of Living Together," *Geoforum* 39, no. 1 (2008): 83–87.

19. Harrison, *Heritage*, 9.

20. Shiloh Krupar, *Hot Spotter's Report: Military Fables of Toxic Waste* (Minneapolis: University of Minnesota Press, 2013), 227.

21. Stuart Hall, "Whose Heritage? Un-settling 'The Heritage,' Re-imagining the Post-nation," *Third Text* 49 (1999–2000): 3–13; David Lowenthal, *The Heritage Crusade and the Spoils of History* (Cambridge: Cambridge University Press, 1998); Harrison, *Heritage*.

22. Laurajane Smith, *Uses of Heritage* (London: Routledge, 2006); Gareth Hoskins, "Vagaries of Value at California State Parks: Towards a Geographical Axiology," *Cultural Geographies* 23, no. 2 (2015): 301–19.

23. Rodney Harrison, "Beyond 'Natural' and 'Cultural' Heritage: Towards an Ontological Politics of Heritage in the Age of the Anthropocene," *Heritage and Society* 8, no. 1 (2015): 24–42.

24. Cornelius Holtorf and Graham Fairclough, "The New Heritage

and Re-shapings of the Past," in *Reclaiming Archaeology: Beyond the Tropes of Modernity*, ed. Alfredo González-Ruibal (London: Routledge, 2013), 197–210.

25. Ian Alden Russell, "Towards an Ethics of Oblivion and Forgetting: The Parallax View," *Heritage and Society* 5, no. 2 (2012): 262.

26. Miriam Clavir, *Preserving What Is Valued: Museums, Conservation, and First Nations* (Vancouver: UBC Press, 2002); Ioannis Poulios, "Moving Beyond a Values-Based Approach to Heritage Conservation," *Conservation and Management of Archaeological Sites* 12, no. 2 (2010): 170–85.

27. Siân Jones, "The Growth of Things and the Fossilisation of Heritage," in *A Future for Archaeology: The Past in the Present*, ed. Robert Layton, Stephen Shennan, and Peter Stone (London: UCL Press, 2006), 113.

28. Joshua S. Martin, N. Adam Smith, and Clinton D. Francis, "Removing the Entropy from the Definition of Entropy: Clarifying the Relationship between Evolution, Entropy, and the Second Law of Thermodynamics," *Evolution: Education and Outreach* 6, no. 30 (2013).

29. Ibid., 5.

30. Don S. Lemons, *A Student's Guide to Entropy* (Cambridge: Cambridge University Press, 2013), 160.

31. Stephen Cairns and Jane M. Jacobs, *Buildings Must Die* (Cambridge, Mass.: MIT Press, 2014), 69.

32. Ibid.

33. Robert Smithson, "Entropy Made Visible," in *Robert Smithson: Collected Writings*, ed. Jack Flam (Berkeley: University of California Press, 1996), 307.

34. Robert Smithson, "Entropy and the New Monuments," in *Robert Smithson: Collected Writings*, 13.

35. Smithson, "Entropy Made Visible," 307.

36. Smithson, "Entropy and the New Monuments," 21.

37. Jeremy Till, *Architecture Depends* (Cambridge, Mass.: MIT Press, 2013), 105.

38. Gavin Lucas, "Time and the Archaeological Archive," *Rethinking History* 14, no. 3 (2010): 355.

39. Yvonne Whelan and Niamh Moore, *Heritage, Memory, and the Politics of Identity: New Perspectives on the Cultural Landscape* (London: Ashgate, 2006); Sharon Macdonald, *Memorylands: Heritage and Identity in Europe Today* (London: Routledge, 2013).

40. Paul Gough, "Sites in the Imagination: The Beaumont Hamel

Newfoundland Memorial on the Somme," *Cultural Geographies* 11, no. 3 (2004): 235–58; Nuala Johnson, "Sculpting Heroic Histories: Celebrating the Centenary of the 1798 Rebellion in Ireland," *Transactions of the Institute of British Geographers* 19, no. 1 (1994): 78–93; Charles Withers, "Place, Memory, Monument: Memorializing the Past in Contemporary Highland Scotland," *Ecumene* 3, no. 3 (1996): 325–44.

41. See, however, David C. Harvey, "Heritage Pasts and Heritage Presents: Temporality, Meaning and the Scope of Heritage Studies," *International Journal of Heritage Studies* 7, no. 4 (2001): 319–38.

42. Mihaly Csikszentmihalyi and Eugene Rochberg-Halton, *The Meaning of Things: Domestic Symbols and the Self* (Cambridge: Cambridge University Press, 1981); Janet Hoskins, *Biographical Objects: How Things Tell the Stories of People's Lives* (New York: Routledge, 1998); Divya Tolia-Kelley, "Locating Processes of Identification: Studying the Precipitates of Re-memory through Artefacts in the British Asian Home," *Transactions of the Institute of British Geographers* 29, no. 3 (2004): 314–29.

43. Aron Vinegar and Jorge Otero-Pailos, "On Preserving the Openness of the Monument," *Future Anterior* 9, no. 2 (2012), iv.

44. Cornelius Holtorf and Oscar Ortman, "Endangerment and the Conservation Ethos in Natural and Cultural Heritage: The Case of Zoos and Archaeological Sites," *International Journal of Heritage Studies* 14, no. 1 (2008), 86.

45. C. Nadia Seremetakis, *The Senses Still: Perception and Memory as Material Culture in Modernity* (Boulder, Colo.: Westview Press, 1994), 9.

46. Bjørnar Olsen, *In Defense of Things: Archaeology and the Ontology of Objects* (New York: AltMira Press, 2013).

47. Þóra Pétursdóttir and Bjørnar Olsen, "Introduction: An Archaeology of Ruins," in Olsen and Pétursdóttir, *Ruin Memories,* 9.

48. Kevin Hetherington, "Spatial Textures: Place, Touch, and Praesentia," *Environment and Planning A* 35, no. 11 (2003): 1933–44; "The Ruin Revisited," in *Trash Cultures: Objects and Obsolescence in Cultural Perspective*, ed. Gillian Pye (Oxford: Peter Lang, 2010), 15–37.

49. It is also worth mentioning that forgetting, in relation to ideas of the self, is largely a fiction, and the absent present may re-emerge as a haunting, the return of the repressed. Steve Pile, *Real Cities: Modernity, Space, and the Phantasmagorias of City Life* (London: Sage, 2005).

50. Aron Vinegar and Jorge Otero-Pailos, "What a Monument Can Do," *Future Anterior* 8, no. 2 (2011): iv.

51. Elizabeth Spelman, *Repair: The Impulse to Restore in a Fragile World* (Boston: Beacon Press, 2003).

52. Giacomo D'Alisa, Federico Demaria, and Giorgos Kallis, eds., *Degrowth: A Vocabulary for a New Era* (London: Routledge, 2014). See also http://dark-mountain.net/ and http://www.friendsofthepleistocene.com/.

53. Ernest Callenbach, "Last Words to an America in Decline," TomDispatch.com, May 6, 2012, http://www.tomdispatch.com/post/175538.

54. Caitlin DeSilvey and Tim Edensor, "Reckoning with Ruins," *Progress in Human Geography* 37, no. 4 (2012): 465–85; Julia Hell and Andreas Schönle, eds., *Ruins of Modernity* (Durham, N.C.: Duke University Press, 2010).

55. Robert Hass, "Meditation at Lagunitas," in *Praise* (New York: Ecco Press, 1979), 4.

56. Terry Eagleton, *The Ideology of the Aesthetic* (Oxford: Blackwell, 1990); Tim Edensor, *Industrial Ruins: Space, Aesthetics, and Materiality* (New York: Berg, 2007).

57. Mark Jackson, "Plastic Islands and Processual Grounds: Ethics, Ontology, and the Matter of Decay," *Cultural Geographies* 20, no. 2 (2013): 207.

58. Jane M. Jacobs, "A Geography of Big Things," *Cultural Geographies* 13, no. 1 (2006): 1–27.

59. Till, *Architecture Depends*; Douglas Murphy, *The Architecture of Failure* (Winchester: Zero Books, 2012); Fred Scott, *On Altering Architecture* (London: Routledge, 2008); Roger Harbison, *The Built, the Unbuilt, and the Unbuildable* (Cambridge, Mass.: MIT Press, 1991); Ed Hollis, *The Secret Lives of Buildings* (London: Portobello Books, 2010); Stewart Brand, *How Buildings Learn* (New York: Penguin, 1994); Cairns and Jacobs, *Buildings Must Die.*

60. Moshen Mostafavi and David Leatherbarrow, *On Weathering: The Life of Buildings in Time* (Cambridge, Mass.: MIT Press, 1999), 16

61. Karen Dale and Gibson Burrell, "Disturbing Structure: Reading the Ruins," *Culture and Organization* 17, no. 2 (2011): 108.

62. Olsen and Pétursdóttir, *Ruin Memories*, 12.

63. Jamie Lorimer, "Multinatural Geographies for the Anthropocene," *Progress in Human Geography* 36, no. 5 (2012): 593–612.

64. Daniela Sandler, "Counterpreservation: Decrepitude and

Memory in Post-unification Berlin," *Third Text* 25, no. 6 (2011): 687–97.

65. Introna, "Ethics and Flesh," 58.

2. Memory's Ecologies

1. Georges Bataille, "Volume II: The History of Eroticism," in *The Accursed Share: An Essay on General Economy* (New York: Zone Books, 1993), 81.

2. Ibid.

3. Mary Douglas, *Purity and Danger* (London: Routledge, 1966), 160.

4. Ibid.

5. Kevin Hetherington, "Secondhandedness: Consumption, Disposal, and Absent Presence," *Society and Space* 22, no. 1 (2004): 157–73.

6. Tim Edensor, "Waste Matter: The Debris of Industrial Ruins and the Disordering of the Material World," *Journal of Material Culture* 10, no. 3 (2005): 318.

7. Douglas, *Purity and Danger,* 160; Sappho, Fragment 84, in *Sappho: A New Translation*, trans. Mary Barnard (Berkeley: University of California Press, 1958).

8. For an exploration of related themes, see Bjørnar Olsen, *In Defence of Things: Archaeology and the Ontology of Objects* (Plymouth, U.K.: AltaMira, 2010).

9. Peter Sloterdijk, *Critique of Cynical Reason* (Minneapolis: University of Minnesota Press, 1987), 151.

10. Martin Jones, "Environmental Archaeology," in *Archaeology: The Key Concepts*, ed. Colin Renfrew and Paul Bahn (London: Routledge, 2005), 85–89.

11. Rebecca Solnit, *Savage Dreams: A Journey into the Landscape Wars of the American West* (Berkeley: University of California Press, 1999), 91.

12. Ian Alden Russell cites Slavoj Žižek's observation that in these moments, "an 'epistemological' shift in the subject's point of view always reflects an 'ontological' shift in the object itself." Ian Alden Russell, "Towards an Ethics of Oblivion and Forgetting: The Parallax View," *Heritage and Society* 5, no. 2 (Fall 2012): 255. See also comments on epistemology and ontology in relation to a similar object in Stephen Jay Gould and Rosamond Wolff Purcell, *Crossing Over: Where Art and Science Meet* (New York: Three Rivers Press, 2000).

13. Michael Taussig, "Miasma," in *Culture and Waste: The Creation and Destruction of Value*, ed. Gay Hawkins and Stephen Muecke (Lanham, Md.: Rowman & Littlefield, 2003), 15–16.

14. Ibid., 16.

15. Arjun Appadurai, ed., *The Social Life of Things: Commodities in Cultural Perspective* (Cambridge: Cambridge University Press, 1986); Victor Buchli, ed., *The Material Culture Reader* (Oxford: Berg, 2002); Mary Beaudry and Dan Hicks, eds., *The Oxford Handbook of Material Culture Studies* (Oxford: Oxford University Press, 2010).

16. Tom Roberts, "From Things to Events: Whitehead and the Materiality of Process," *Environment and Planning D: Society and Space* 32, no. 6 (2014): 968–83.

17. Rudi Colloredo-Mansfeld, "Matter Unbound," *Journal of Material Culture* 8, no. 33 (2004): 246, 250.

18. Hetherington, "Secondhandedness"; Gavin Lucas, "Disposability and Dispossession in the Twentieth Century," *Journal of Material Culture* 7, no. 1 (2002): 5–22; Melanie Van der Hoorn, "Exorcizing Remains: Architectural Fragments as Intermediaries between History and Individual Experience," *Journal of Material Culture* 8, no. 2 (2003): 189–213; Nicky Gregson, Alan Metcalfe, and Louise Crewe, "Identity, Mobility and the Throwaway Society," *Environment and Planning D: Society and Space* 25 (2007): 682–700.

19. Tim Edensor, *Industrial Ruins: Aesthetics, Materiality, and Memory* (Oxford: Berg, 2005), 100.

20. Tim Ingold, "Bringing Things to Life: Creative Entanglements in a World of Materials," NCRM Working Paper (Realities/Morgan Centre, University of Manchester, 2010), 2–3, http://eprints.ncrm.ac.uk/1306/.

21. Brian Neville and Johanne Villeneuve, "Introduction: In Lieu of Waste," in *Waste-Site Stories: The Recycling of Memory,* ed. Brian Neville and Johanne Villeneuve (Albany: State University of New York Press, 2002), 2.

22. Walter Benjamin, *The Arcades Project*, trans. H. Eiland and K. McLaughlin (Cambridge, Mass.: Belknap Press, 1999); Julia Hell and Andreas Schönle, eds., *Ruins of Modernity* (Durham, N.C.: Duke University Press, 2010).

23. Edward S. Casey, *Remembering: A Phenomenological Study* (Bloomington: Indiana University Press, 2000), 311.

24. Mike Pearson and Michael Shanks, *Theatre/Archaeology* (London: Routledge, 2001), 158.

25. Chris Ford, personal communication, Grant-Kohrs Ranch, Deer Lodge, Montana, September 10, 2002.
26. Buchli, *Material Culture Reader*, 15.
27. Dydia DeLyser, "Authenticity on the Ground: Engaging the Past in a California Ghost Town," *Annals of the Association of American Geographers* 89 (1999): 602–32.
28. Susanne Küchler, "The Place of Memory," in *The Art of Forgetting*, ed. Adrian Forty and Susanne Küchler (Oxford: Berg, 1999), 57, 62.
29. Ibid., 63.
30. James W. Feldman, *A Storied Wilderness: Rewilding the Apostle Islands* (Seattle: University of Washington Press, 2011).
31. Mieke Bal, Jonathan Crewe, and Leo Spitzer, eds., *Acts of Memory: Cultural Recall in the Present* (Hanover, N.H.: University Press of New England, 1999), vii.
32. David Gross, "Objects from the Past," in Neville and Villeneuve, *Waste-Site Stories*, 36.
33. Küchler, "Place of Memory," 59.
34. Christopher Woodward, *In Ruins* (London: Vintage, 2002).
35. Miles Ogborn, "Archives," in *Patterned Ground: Entanglements of Nature and Culture*, ed. Stephan Harrison, Steve Pile, and Nigel Thrift (London: Reaktion, 2004), 240.
36. Steve Hinchcliffe, "Inhabiting: Landscapes and Natures," in *Handbook of Cultural Geography*, ed. Kay Anderson, Mona Domosh, Steve Pile, and Nigel Thrift (London: Sage, 2003), 207–25; Tim Ingold, *The Perception of the Environment: Essays in Livelihood, Dwelling, and Skill* (London: Routledge, 2000); Elizabeth Hallam and Tim Ingold, eds., *Making and Growing: Anthropological Studies of Organisms and Artefacts* (London: Ashgate, 2014).
37. Edensor, *Industrial Ruins*, 122.
38. Phil Dunham, "Dust," in Harrison, Pile, and Thirft, *Patterned Ground*, 100.
39. Gay Hawkins and Stephen Muecke, "Introduction: Cultural Economies of Waste," in Hawkins and Muecke, *Culture and Waste*, 7.
40. Victor Buchli and Gavin Lucas, *Archaeologies of the Contemporary Past* (London: Routledge, 2001), 5.
41. Tristan Tzara, *Seven Dada Manifestos and Lampisteries* (London: Calder Publications, 1992), 39.
42. Sarah Whatmore, "Generating Materials," in *Using Social Theory*, ed. Michael Pryke, Gillian Rose, and Sarah Whatmore (London: Sage, 2003), 98.

43. Barbara Kirshenblatt-Gimblett, *Destination Culture: Tourism, Museums, and Heritage* (Berkeley: University of California Press, 1998), 168.
44. Buchli and Lucas, *Archaeologies*.
45. North Missoula Community Development Corporation, "Moon Randolph Homestead Strategic Plan Update: 2015–2024," unpublished report, Missoula, Mont.

3. When Story Meets the Storm

1. S. A. Chang, "Continental Shift," *Preservation*, September–October 2006, 22–29.
2. National Trust, *Shifting Shores: Living with a Changing Coastline* (London: National Trust, 2005); "Coastal Policy," *National Trust*, March 2006, http://nationaltrust.org.uk.
3. "Trust Announces Conclusion of Harbour Study," National Trust media release, March 22, 2006, *Objective One*, http://www.objectiveone.com.
4. Royal Haskoning, "Cornwall and Isles of Scilly Shoreline Management Plan Review," 2011, *Cornwall and Isles of Scilly Coastal Advisory Group*, http://www.ciscag.org/index.html.
5. Caitlin DeSilvey, "Making Sense of Transience: An Anticipatory History," *Cultural Geographies* 19, no. 2 (2012): 31–54.
6. Ibid., 35.
7. T. S. Eliot, "Four Quartets: Little Gidding," in *T. S. Eliot: The Complete Poems and Plays, 1909–1950* (New York: Harcourt, Brace, and World, 1962), 144.
8. DeSilvey, "Making Sense," 35.
9. Eliot, "Four Quartets: Little Gidding," 139.
10. Michael Taussig, *Walter Benjamin's Grave* (Chicago: University of Chicago Press, 2006), vii.
11. Tim Edensor, "Vital Urban Materiality and Its Multiple Absences: The Building Stone of Central Manchester," *Cultural Geographies* 20, no. 4 (2013): 447–65.
12. Robert Felce, *A History of Mullion Cove, Cornwall* (Mullion: Westcountry Printing and Publishing, 2012).
13. Months later it was revealed that English Heritage had been consulted in 2012 about a proposal to use concrete in harbor repairs and had advised the Cornwall Council conservation officer at the time that English Heritage would not support the use of concrete for reconstruction. This advice was apparently never communicated to the National Trust.

14. Robert Felce, personal communication, November 21, 2013.
15. Robert Felce, "History of Mullion Cove and Harbour," https://sites.google.com/site/historyofmullioncoveandharbour.
16. Felce, *History of Mullion Cove*, 117.
17. Emilie Cameron, "New Geographies of Story and Story-telling," *Progress in Human Geography* 36, no. 5 (2012): 573–92.
18. Eliot, "Four Quartets: The Dry Salvages," 133.
19. Richard Smith, "The Mad Dad," *Daily Mirror*, January 4, 2014.
20. Felce, *History of Mullion Cove*, 13. The Daniel J. Draper lifeboat began its service in September 1867.
21. Felce, "History of Mullion Cove and Harbour," January 4, 2014.
22. BBC Cornwall Radio Cornwall, *Breakfast Programme*, January 8, 2014.
23. Mullion Parish Council, minutes of Mullion Parish council meeting, January 21, 2014, *Mullion Parish Council*, http://www.mullionparishcouncil.org.uk/meetings.html.
24. W. B. Graeme, "National Trust to Probe Mullion Cove Harbour Damage," *West Briton*, January 17, 2014.
25. BBC Cornwall Radio Cornwall, *Breakfast Programme*, February 5, 2014.
26. Ibid., February 16, 2014.
27. Eliot, "Four Quartets: East Coker," 126.
28. Lizard National Trust, *Facebook*, February 24, 2014, https://www.facebook.com/LizardNT.
29. National Trust, *Mullion Cove: A Strategy for Coping with Climate Change* (Lanhydrock, Bodmin, U.K.: National Trust, 2006).
30. Anne Michaels, *The Winter Vault* (London: Bloomsbury, 2010), 275.
31. Paul Basu, "Cairns in the Landscape: Migrant Stones and Migrant Stories in Scotland and Its Diaspora," in *Landscapes beyond Land: Routes, Aesthetics, Narratives*, ed. Arnar Arnason, Nicholas Ellison, Jo Vergunst, and Andrew Whitehouse (New York: Berghan Books, 2012), 135. See also Avril Maddrell, "A Place for Grief and Belief: The Witness Cairn, Isle of Whithorn, Galloway, Scotland," *Social and Cultural Geography* 10, no. 6 (2009): 675–93.
32. Judith Butler, *Precarious Life: The Powers of Mourning and Violence* (London: Verso, 2006), xvii.
33. Aron Vinegar and Jorge Otero-Pailos, "On Preserving the Openness of the Monument," *Future Anterior* 9, no. 2 (2012): 2.
34. Eliot, "Four Quartets: Little Gidding," 142.

35. Planning applications PA14/05089 and PA14/05090, *Cornwall Council*, http://planning.cornwall.gov.uk/online-applications/.
36. Ibid., PA14/05090; English Heritage comment, July 21, 2014.
37. "Search the List," *Historic England*, http://list.english-heritage.org.uk, list entry 1158181.
38. Eliot, "Four Quartets: Little Gidding," 142.
39. Eliot, "Four Quartets: Burnt Norton," 121.

4. Orderly Decay

1. Wayne Cocroft and Magnus Alexander, *Atomic Weapons Research Establishment, Orford Ness, Suffolk, Survey Report*, Research Department Report Series No. 10-2009 (Swindon: English Heritage, 2009).
2. National Trust, *Orford Ness Guidebook* (Swindon: National Trust, 2003).
3. Paddy Heazel, *Most Secret: The Hidden History of Orford Ness* (Stroud: History Press, 2010), 181.
4. Ibid.
5. W. G. Sebald, *Rings of Saturn* (London: Harvill, 1998), 237.
6. Angus Wainwright, "Orford Ness: A Landscape in Conflict?," in *Europe's Deadly Century: Perspectives on 20th Century Conflict Heritage*, ed. N. Forbes, R. Page, and G. Perez (Swindon: English Heritage, 2009), 136.
7. Ibid., 140.
8. National Trust, n.d., "Orford Ness: A Statement of Significance," sourced from Orford Ness property archives.
9. Wainwright, "Orford Ness," 140.
10. Ibid.
11. Charles Merewether, "Traces of Loss," in *Irresistible Decay*, ed. Michael Roth, Claire Lyons, and Charles Merewether (Los Angeles: Getty Research Institute, 1997), 33.
12. Cocroft and Alexander, *Atomic Weapons*, 50.
13. National Trust, *Orford Ness Guidebook,* 23.
14. Jack Watkins, "Orford Ness No Longer an 'Awful Mess,'" *Telegraph,* August 20, 2009, http://www.telegraph.co.uk/.
15. Alois Riegl, "The Modern Cult of Monuments," in *Historical and Philosophical Issues in the Conservation of Cultural Heritage*, ed. N. S. Price, M. K. Talley, and A. M. Vaccaro (1903; repr., Los Angeles: Getty Conservation Institute,1996), 69–83.
16. Ibid., 78.

17. Ibid., 75.
18. Ibid., 73.
19. Ibid., 74.
20. Ibid., 73.
21. David Lowenthal, "Material Preservation and Its Alternatives," *Perspecta* 25 (1989): 72.
22. Cornelius Holtorf, "On Pastness: A Reconsideration of Materiality in Archaeological Object Authenticity," *Anthropological Quarterly* 86, no. 2 (2013): 427–43.
23. Stephen Cairns and Jane M. Jacobs, *Buildings Must Die: A Perverse View of Architecture* (Cambridge, Mass.: MIT Press, 2014), 71.
24. Riegl, "Modern Cult," 73.
25. In the pre–World War I era, the legislative context in Germany was somewhat different from that in the United Kingdom in that responsibility for heritage protection lay with federal states rather than the national government. See Rudy J. Koshar, "On Cults and Cultists: German Historic Preservation in the Twentieth Century," in *Giving Preservation a History: Histories of Historic Preservation in the United States*, ed. Max Page and Randall Mason (New York: Routledge, 2004), 45–78.
26. National Trust, "LIFE Projects on Orford Ness," *National Trust*, http://www.nationaltrust.org.uk/orford-ness-national-nature-reserve/features/life-projects-on-orford-ness.
27. Richard Mabey, *Flora Britannica: The Concise Edition* (London: Chatto and Windus, 1998), 178.
28. John Hersey, *Hiroshima* (1946; repr., New York: Bantam Books, 1986), 69.
29. National Trust, "Orford Ness: A Statement of Significance."
30. John Beck, "Concrete Ambivalence: Inside the Bunker Complex," *Cultural Politics* 7 (2011): 79–102, 82; Peter Coates, Tim Cole, Marina Dudley, and Chris Pearson, "Defending Nation, Defending Nature? Militarized Landscapes and Military Environmentalism in Britain, France, and the United States," *Environmental History* 16 (2011): 456–91.
31. Jeffery Sasha Davis, "Military Natures: Militarism and the Environment," *GeoJournal* 69 (2007): 131.
32. David Havlick, "Logics of Change for Military-to-Wildlife Conversions in the United States," *GeoJournal* 69 (2007): 151–64; Coates et al., "Defending Nation."
33. Sophia Davis, "Military Landscapes and Secret Science: The Case of Orford Ness," *Cultural Geographies* 15, no. 1 (2008): 143–49.

34. Jeremy Musson, cited in Christopher Woodward, *In Ruins* (London: Vintage, 2001), 223.

35. For further discussion of artistic engagements with Orford Ness, see Louise K. Wilson, "Processional Engagement: Sebaldian Pilgrimages to the Ness," RGS-IBG Annual Conference, London, session on "Cold War Bunkers: Exceptionalism, Affect, Materiality and Aftermath," August 2014; and https://www.nationaltrust.org.uk/orford-ness-national-nature-reserve/features/previous-art-projects-on-orford-ness/.

36. Louise K. Wilson, "Notes on a Record of Fear: On the Threshold of the Audible," in *Contemporary Archaeologies: Excavating Now*, ed. Cornelius Holtorf and Angela Piccini (Frankfurt am Main: Peter Lang, 2009), 117.

37. Other works in the series were installed at Pripyat, the abandoned town adjacent to Chernobyl, and an antiquarian bookshop in London.

38. "Orford Ness Pebble Turned into Giant Artworks for Aldeburgh Festival," *BBC News*, June 23, 2014, http://www.bbc.com/.

39. Matthew Flintham, "The Military–Pastoral Complex: Contemporary Representations of Militarism in the Landscape," *Tate Papers* 17 (2012), http://www.tate.org.uk/research/publications/tate-papers/.

40. Wainwright, "Orford Ness," 136.

41. David Lowenthal, "The Value of Age and Decay," in *Durability and Change: The Science, Responsibility, and Cost of Sustaining Cultural Heritage*, ed. W. E. Krumbein, P. Brimblecombe, D. E. Cosgrove, and S. Staniforth (Chichester: Wiley, 1994), 40.

42. Caitlin DeSilvey, "Palliative Curation: Art and Entropy on Orford Ness," in *Ruin Memories: Materialities, Aesthetics, and the Archaeology of the Recent Past*, ed. Bjørnar Olsen and Þóra Pétursdóttir (London: Routledge, 2014), 79–91.

43. M. L. Rozenweig, "Reconciliation Ecology and the Future of Species Diversity," *Oryx* 37 (2003): 197–205; Robert A. Francis and Jamie Lorimer, "Urban Reconciliation Ecology," *Journal of Environmental Management* 92 (2011): 1429–37.

44. J. B. Jackson, *The Necessity for Ruins* (Amherst: University of Massachusetts Press, 1980).

45. Wayne Cocroft and R. J. C. Thomas, *Cold War: Building for Nuclear Confrontation, 1946–1989* (Swindon: English Heritage, 2003).

46. Luke Bennett, "Bunkerology: A Case Study in the Theory and Practice of Urban Exploration," *Environment and Planning D:*

Society and Space 29 (2011): 421–34; Ian Strange and Ed Walley, "Cold War Heritage and the Conservation of Military Remains in Yorkshire," *International Journal of Heritage Studies* 13 (2007): 154–69; John Schofield and Wayne Cocroft, *A Fearsome Heritage: Diverse Legacies of the Cold War* (Walnut Creek, Calif.: Left Coast Press, 2009).

47. Cocroft and Alexander, *Atomic Weapons.*

48. For a discussion of another debate that exposed the paradox generated when attempts are made to apply conventional conservation designations and protections to a site which is valued for its enrollment in ongoing processes of decay, see Mats Burström, "Garbage or Heritage: The Existential Dimension of a Car Cemetery," in Holtorf and Piccini, *Contemporary Archaeologies*, 131–43.

49. Wainwright, "Orford Ness,"140.

50. Riegl, "Modern Cult," 77–78.

51. Ibid., 76

52. Personal communication with Grant Lohoar, Orford Ness, March 29, 2012.

53. Wainwright, "Orford Ness," 141.

54. Michael Roth, "Irresistible Decay: Ruins Reclaimed," in Roth, Lyons, and Merewether, *Irresistible Decay*, 2.

55. Lowenthal, "Value of Age and Decay," 47.

56. John Ruskin, "The Lamp of Memory: II," in *Historical and Philosophical Issues in the Conservation of Cultural Heritage*, ed. N. S. Price, M. K. Talley, and A. M. Vaccaro (1849; repr., Los Angeles: Getty Conservation Institute, 1996), 322–23.

57. Riegl, "Modern Cult," 74.

58. Ibid.

59. Ibid.

60. Ibid.

61. Ibid., 77.

5. A Positive Passivity

1. Colin D. Meurk, "Recombinant Ecology of Urban Areas: Characterisation, Context, and Creativity," in *The Routledge Handbook of Urban Ecology*, ed. Ian Douglas, David Goode, Mike Houck, and Rusong Wang (London: Routledge, 2010), 198–220.

2. Definition cited in Jamie Lorimer and Clemens Driessen, "Wild Experiments: Rethinking Environmentalism in the Anthropo-

cene," *Transactions of the Institute of British Geographers* 39, no. 2 (2014): 170.

3. Detlef Lange, ed., *The Ruhrgebiet Industrial Heritage Atlas* (Essen: Regionalverband Ruhr, 2004).

4. NVA, "Statement of Intent," in *To Have and to Hold*, ed. Gerrie van Noord (Edinburgh: Luath Press, 2011), 22.

5. Orvar Löfgren, "Wear and Tear," *Ethnologia Europea* 35, no. 1–2 (2005): 53.

6. John Piper, *Buildings and Prospects* (London: Architectural Press, 1948).

7. Lesley Head and Jennifer Atchison, "Cultural Ecology: Emerging Plant–Human Geographies," *Progress in Human Geography* 33, no. 2 (2009): 236–45; Jennifer Atchison and Lesley Head, "Eradicating Bodies in Invasive Plant Management," *Environment and Planning D: Society and Space* 31 (2013): 951–68; Hannah Pitt, "On Showing and Being Shown Plants: A Guide to Methods for More-than-Human Geography," *Area* 47, no. 1 (2014): 48–55.

8. Michael Pollan, *Second Nature: A Gardener's Education* (New York: Dell, 1991). See also Russell Hitchings, "Expertise and Inability: Cultured Materials and the Reason for Some Retreating Lawns in London," *Journal of Material Culture* 11, no. 3 (2006), 364–81.

9. Peter Latz, "Landscape Park Duisburg Nord: The Metamorphosis of an Industrial Site," in *Manufactured Sites: Rethinking the Post-industrial Landscape*, ed. Niall Kirkwood (Abingdon: Taylor & Francis, 2001), 151.

10. Ibid., 153.

11. Ibid., 158.

12. Elissa Rosenberg, "Gardens, Landscape, Nature: Duisburg Nord, Germany," in *The Hand and the Soul: Aesthetics and Ethics in Architecture and Art* (Charlottesville: University of Virginia Press, 2009), 216.

13. Latz, "Landscape Park Duisburg Nord," 159.

14. Georg Simmel, "The Ruin," in *Essays on Sociology, Philosophy, and Aesthetics*, ed. Kurt H. Wolf (New York: Harper & Row, 1965), 260.

15. Kevin Hetherington, "The Ruin Revisited," in *Trash Culture: Objects and Obsolescence in Cultural Perspective*, ed. Gillian Pye (Oxford: Peter Lang, 2010), 20.

16. Hitchings, "Expertise and Inability," 368.

17. My time at Duisburg Nord in July 2012 was limited. Over two days, I gathered the observations and impressions I share here; immediately afterward, I traveled to Bayreuth University, where I presented the work to Matthew Hannah and his students. I benefited greatly from their critical comments on my initial attempts to make sense of my experience.

18. Rosenberg, "Gardens, Landscape, Nature."

19. Jozef Keulartz, "The Different Faces of History in Postindustrial Landscapes," paper presented at the Society for Ecological Restoration Fifth World Congress, October 9, 2013, Madison, Wis.

20. Kerstin Barndt, "'Memory Traces of an Abandoned Set of Futures': Industrial Ruins in the Postindustrial Landscapes of Germany," in *Ruins of Modernity*, ed. Julia Hell and Andreas Schönle (Durham, N.H.: Duke University Press), 281.

21. Ibid.

22. ThyssenKrupp, http://www.thyssenkrupp.com/.

23. Latz, "Landscape Park Duisburg Nord," 151.

24. Rosenberg, "Gardens, Landscape, Nature," 225.

25. Tilman Latz, "Once Upon a Time . . . ," in van Noord, *To Have and to Hold*, 67.

26. The identical definition is also found in Lange's *Ruhrgebiet Industrial Heritage Atlas*, 86.

27. Information about each site included in the *Route Industrienatur* is available (in German) at *Metropoleruhr*, http://www.metropoleruhr.de/freizeit-sport/natur-erleben/route-industrienatur.html.

28. Richard J. Hobbs, Eric S. Higgs, and Carol M. Hall, *Novel Ecosystems: Intervening in the New Ecological World Order* (Chichester: Wiley-Blackwell, 2013), 58.

29. "Route Industrienatur," *Metropoleruhr*.

30. Fred Pearce, *The New Wild: Why Invasive Species Will Be Nature's Salvation* (Boston: Beacon Press, 2015).

31. Rosenberg, "Gardens, Landscape, Nature," 210.

32. Shiloh Krupar, *Hot Spotter's Report: Military Fables of Toxic Waste* (Minneapolis: University of Minnesota Press, 2013), 223.

33. Hitchings, "Expertise and Inability," 377.

34. Emma Marris, *Rambunctious Garden: Saving Nature in a Postwild World* (New York: Bloomsbury, 2011).

35. Dan Swanton, "Afterimages of Steel: Dortmund," *Space and Culture* 15, no. 4 (2012): 276.

36. Torgeir Bangstad, "Defamiliarization, Conflict, and Authentic-

ity: Industrial Heritage and the Problem of Representation," Ph.D. thesis, NTNU-Trondheim, Norway, 2014.

37. Kerstin Barndt, "Layers of Time: Industrial Ruins and Exhibitionary Temporalities," *PMLA* 125, no. 1 (2010): 8.

38. Simmel, "Ruin," 261.

39. Ibid., 263.

40. For a sympathetic analysis of Duisburg Nord and a discussion of the origins of the term *Industrienatur*, see Anna Storm, *Post-industrial Landscape Scars* (New York: Palgrave Macmillan, 2014).

41. *The Invisible College*, http://www.theinvisiblecollege.org.uk/locale.

42. I first visited Kilmahew/St. Peter's when I was invited by Hayden Lorimer and Mike Gallagher to participate in an Invisible College event in September 2012. Many of my observations in this chapter draw on conversations and presentations associated with this initial exposure. I returned to the site on an informal visit with Mike, Hayden, and Erin Despard in May 2014.

43. NVA, "Statement of Intent," 22.

44. Ed Hollis, "Anxious Care and Unsightly Aids," in van Noord, *To Have and to Hold*, 54.

45. "Kilmahew/St. Peter's," video available at *NVA Public Art*, http://nva.org.uk/artwork/kilmahew-st-peters/.

46. A record of the workshop event is available at "Workshops," *The Invisible College*, http://www.theinvisiblecollege.org.uk/workshops; and Michael Gallagher, "Sounding Ruins: Reflections on the Production of an 'Audio Drift,'" *Cultural Geographies* 22, no. 3 (2015): 467–85.

47. Avanti Architects, "Kilmahew/St. Peter's Masterplan Summary Document (2011)," 6, *NVA Public Art*, http://nva.org.uk/artwork/kilmahew-st-peters/#tab-2.

48. Ibid., 6.

49. Ibid., 20–21.

50. Ibid., 7.

51. Ibid., 33.

52. Ibid., 23.

53. Hayden Lorimer and Michael Gallagher, "Ruination and Re-invention: A Self-Guided Walk around Kilmahew Estate near Cardross," *Royal Geographical Society with IBG*, "Discovering Britain," http://www.discoveringbritain.org/walks/region/scotland/kilmahew.html.

54. NVA, "Statement of Intent," 21; Lorimer and Gallagher, "Ruination and Reinvention."
55. Woodland Trust, "Interactive Map," *Ancient Tree Hunt*, http://www.ancient-tree-hunt.org.uk/discoveries/interactivemap/.
56. Avanti Architects, "Kilmahew/St. Peter's Masterplan," 18.
57. Ibid.
58. Simmel, "Ruin," 263.
59. Barnabas Calder, The Invisible College public talk, September 8, 2012, The Lighthouse, Glasgow.

6. Boundary Work

1. J. B. Jackson, *The Necessity for Ruins* (Amherst: University of Massachusetts Press, 1980).
2. Cindi Katz, "Whose Nature, Whose Culture? Private Productions of Space and the 'Preservation' of Nature," in *Remaking Reality: Nature at the Millennium*, ed. Bruce Braun and Noel Castree (London: Routledge, 1998), 54.
3. Cornelius Holtorf, "Averting Loss in Aversion in Cultural Heritage," *International Journal of Heritage Studies* 21, no. 4 (2014): 405–21. See also David Gibbs, Aidan While, and Andrew Jonas, "Governing Nature Conservation: The European Union Habitats Directive and Conflict around Estuary Management," *Environment and Planning A* 39 (2007): 339–58.
4. Sara B. Pritchard, "Joining Environmental History with Science and Technology Studies: Promises, Challenges, and Contributions," in *New Natures*, ed. Dolly Jørgensen, Finn Arne Jørgensen, and Sara B. Pritchard (Pittsburgh: University of Pittsburgh Press, 2013), 13.
5. Holtorf, "Averting Loss," 8.
6. Alan Holland and Kate Rawles, "The Ethics of Conservation," report presented to the Countryside Council for Wales, Thingmount Series No.1. (Lancaster: Lancaster University Department of Philosophy, 1994); Helen Ghosh, "From the Director-General," *National Trust Magazine*, Autumn 2014: 16.
7. Holland and Rawles, "Ethics."
8. Gustavo F. Araoz, "Heritage Classifications and the Need to Adjust Them to Emerging Paradigms: The United States Experience," in *Values and Criteria in Heritage Conservation*, ed. A. Tomaszewski (Florence: Edizioni Polistampa, 2008), 167–82, cited in Holtorf, "Averting Loss," 8.
9. Stephen T. Jackson and Richard J. Hobbs, "Ecological Resto-

ration in the Light of Ecological History," *Science* 325 (2009): 568.

10. Rodney Harrison, "Beyond Natural/Cultural Heritage: Towards an Ontological Politics of Heritage in the Age of Anthropocene," *Heritage and Society* 8, no. 1 (2015): 24–42.

11. Cornelius Holtorf and Graham Fairclough, "The New Heritage and Re-shapings of the Past," in *Reclaiming Archaeology: Beyond the Tropes of Modernity*, ed. Alfredo Gonzálz-Ruibal (London: Routledge, 2013), 197–210.

12. Adrian Spalding, Stephen Hartgroves, John Macadam, and David Owens, eds., *The Conservation Value of Abandoned Pits and Quarries* (Truro, Cornwall: Historic Environment Service, Cornwall County Council, 1999).

13. Some experts estimate that up to 25 percent of all SSSIs in the United Kingdom are on former mineral workings; Pete Whitbread-Abrutat, personal communication, September 26, 2014. See also "The Ecological Value of Metalliferous Mining Sites," 2011, *Cornish Mining World Heritage*, http://www.cornish-mining.org.uk/sites/default/files/04%20-%20The%20ecological%20value%20of%20metalliferous%20mining%20sites.pdf.

14. Relevant legislation includes the U.K. Wildlife and Countryside Act 1981, the Town and Country Planning Act 1947, and the Planning (Listed Buildings and Conservation Areas) Act 1990.

15. Ainsley Cocks, "Conserving Mining History to Commence at Wheal Busy," *West Briton*, February 20, 2014, http://www.westbriton.co.uk/Conserving-mining-history-commence-Wheal-Busy/story-20665769-detail/story.html.

16. The conservation work was undertaken by Natural England in partnership with the Tregothnan Estate and the landowner, with the works funded through the Natural England Higher Level Stewardship (HLS) scheme, Historical and Archaeological Feature Protection (HAP).

17. Cocks, "Conserving Mining History."

18. Will Rowland, comment on Cocks, "Conserving Mining History," May 27, 2014.

19. David Hazlehurst, personal communication, June 4, 2015.

20. Shaun Lewin, Falmouth Convention, Field Trip 3: Hydroplutonic Kernow, May 21, 2009.

21. J. R. Smith, *Kennall Vale Archaeological Report* (Truro: Cornwall Trust for Nature Conservation, 1986).

22. Peter Herring, *Kennall Vale Reserve Archaeological Assessment: A*

Report to the Cornwall Wildlife Trust (Truro: Cornwall Archaeological Unit, Cornwall County Council, 1999).

23. S. Adams, "Management and Monitoring Plan: Kennall Vale Nature Reserve, 2001–2006: A Report to Cornwall Wildlife Trust," May 2000.

24. Ann Preston-Jones, *Kennall Vale, Ponsanooth, Cornwall: Archaeological Management and Interpretation* (Truro: Cornwall Council Historic Environment Projects, 2011).

25. Ibid., 17.

26. Robin Kent, "Thirlwall Castle: The Use of Soft Capping in Conserving Ruined Ancient Monuments," *Journal of Architectural Conservation* 19, no. 1 (2013): 35–48; Chris Wood, "Soft Capping: Justifying a Return to the Picturesque," *Context* 90 (2005): 22–24. Other sites where this approach has been applied in the United Kingdom include Wigmore Castle, Jervaulx Abbey, Fountain's Abbey, and Tintagel.

27. Preston-Jones, *Kennall Vale*, 20.

28. Nick Marriott, personal communication, October 6, 2014.

29. Rachel Thomas, "Historic Ruins and Nature Conservation: Towards a Holistic Approach," paper presented at the Ruins Conference, Bath, England, June 2005.

30. James Feldman, *A Storied Wilderness: Rewilding the Apostle Islands* (Seattle: University of Washington Press, 2011), 229.

31. Ibid., 231.

32. "Arrested decay" is the standard preservation treatment for abandoned settlements in the American West. Charles Bovey of Virginia City, Montana, was an early proponent of what he described as a "suspended state of deterioration." The labor required to maintain a structure in a state of apparent near collapse is significant, however, as detailed by Dydia DeLyser in "Authenticity on the Ground: Engaging the Past in a California Ghost Town," *Annals of the Association of American Geographers* 89, no. 4 (1999): 602–32.

33. Tom Wessells, *Forest Forensics: A Field Guide to Reading the Forested Landscape* (New York: Countryman Press, 2010).

34. Bureau of Land Management, *8110—Identifying and Evaluating Cultural Resources* (2004), http://www.blm.gov/pgdata/etc/medialib/blm/wo/Information_Resources_Management/policy/blm_manual.Par.23101.File.dat/8110.pdf.

35. Matt Thompson, "I See a Darkness: What Happens if We Choose Not to Save Historic Structures," unpublished paper, 2014.

36. Rodney Harrison, "Excavating Second Life: Cyber-archaeologies, Heritage, and Virtual Communities," *Journal of Material Culture* 14, no. 1 (2009): 75–106.

37. Australian Antarctic Division, *Heard Island and McDonald Islands Reserve Management Plan* (Canberra: Department of Environment and Heritage, Australian Government, 2005).

38. Rebecca Solnit, "Ghost River: On Photography in Yosemite," in *Yosemite in Time*, ed. Mark Klett, Rebecca Solnit, and Byron Wolfe (San Antonio, Tex.: Trinity University Press, 2005): 28.

39. Katarina Saltzman, "Composting," *Journal of European Ethnology* 35, no. 1–2 (2005): 63.

40. Ibid.

41. Ibid., 68.

42. James Edgar, "'Bats Matter More than Worshippers,' Peer Tells Lords," *Telegraph,* June 12, 2014, http://www.telegraph.co.uk/.

43. David Bullock and Jacky Ferneyhough, *When Nature Moves In: A Guide to Managing Wildlife In and Around Buildings* (Swindon: National Trust, 2013).

44. Ron Porley, "Threatened Bryophytes: *Leptodontium gemmascens*," *Field Bryology* 96 (2008): 14–25.

45. Pritchard, "Joining Environmental History," 14.

46. Steve Hinchliffe, "Reconstituting Nature Conservation: Towards a Careful Political Ecology," *Geoforum* 39, no. 1 (2008): 88–97.

47. David Lowenthal, "Natural and Cultural Heritage," *International Journal of Heritage Studies* 11, no. 1 (2005): 81–92.

48. Elizabeth Hallam and Tim Ingold, eds., *Making and Growing: Anthropological Studies of Organisms and Artefacts* (London: Ashgate, 2014).

7. Palliative Curation

1. Details drawn from the English Heritage List Entry Summary, entry 1392631, Orfordness Lighthouse (the full entry is available at http://list.english-heritage.org.uk) and from personal correspondence with Duncan Kent, June 27, 2015.

2. Suffolk Coastal District Council, Waveney District Council, and Environment Agency, "Shoreline Management Plan 7 (Previously Sub-Cell 3C)," May 2015, http://suffolksmp2.org.uk/.

3. Rose Macaulay, *The Pleasure of Ruins* (London: Thames and Hudson, 1953), 23. Middas Dekkers explored this theme exhaustively in *The Way of All Flesh: A Celebration of Decay* (London: Harvill Press, 2000).

4. I returned to Orford Ness in July 2012 and met with Liz Ferretti, the local writer who was leading the Orfordness Lighthouse Project. At that stage, the project was awaiting news of a bid to the Heritage Lottery Fund, which was subsequently granted. I was also involved, at a distance, in a workshop that was held on Orford Ness in June 2012, hosted by the National Trust for local property staff and external stakeholders. The workshop explored strategies for engaging people with coastal change. It introduced the concept of anticipatory history (developed at Mullion) as a potential interpretive tool.

5. Phil Dyke, personal communication , June 4, 2015.

6. "Orfordness Lighthouse Gets Switched Off and Left to the Sea," *BBC News*, June 28, 2013, http://www.bbc.co.uk/. The lighthouse is commonly referred to by the name "Orfordness" and the surrounding spit as "Orford Ness."

7. Martin Fletcher, "Slipping Away: Row Threatens Centuries-Old Lighthouse," *Telegraph*, January 12, 2014, http://www.telegraph.co.uk/.

8. *Orfordness Lighthouse Company*, http://www.orford.org.uk/community/orfordness-lighthouse-company-2/.

9. Suffolk Coastal District Council, planning application DC/14/0206/FUL, available at *East Suffolk: Suffolk Coastal and Waveney Councils*, http://www.suffolkcoastal.gov.uk/.

10. "The National Trust's Position on the Orford Ness Lighthouse," *National Trust*, January 12, 2014, http://ntpressoffice.wordpress.com/2014/01/12/national-trusts-position-on-the-orford-ness-lighthouse/. Natural England, which holds the authority for determining compliance with protective designations, emphasized the need to avoid impact on SSSI and SAC features but supported the proposal as a temporary measure; comments on planning application DC/14/0206/FUL, February 14, 2014.

11. National Trust, *East of England*, February 3, 2014, https://eastofenglandnt.wordpress.com/2014/02/03/orford-ness-lighthouse/.

12. Terry Underwood, comments on planning application DC/14/0206/FUL, February 15, 2014.

13. David Lowenthal, "The Value of Age and Decay," in *Durability and Change: The Science, Responsibility, and Cost of Sustaining Cultural Heritage*, ed. W. E. Krumbein, P. Brimblecombe, D. E. Cosgrove, and S. Staniforth (Chichester: Wiley, 1994), 43.

14. Michael Shanks, "The Life of an Artifact in Interpretive Archaeology," *Fennoscandia Archaeologica* 15 (1998): 17.

15. Ibid.
16. Ibid., 18.
17. Ibid., 19.
18. Ibid., 22.
19. Pepe Romanillos, "Geography, Death, and Finitude," *Environment and Planning A* 43 (2011): 2549.
20. Neil Harris, *Building Lives: Constructing Rites and Passages* (New Haven, Conn.: Yale University Press, 1999), 117.
21. Ibid., 157. See also Stephen Cairns and Jane M. Jacobs, *Buildings Must Die: A Perverse View of Architecture* (Cambridge, Mass.: MIT Press, 2014).
22. Elizabeth Edwards, *The Camera as Historian: Amateur Photographers and Historical Imagination, 1885–1918* (Durham, N.C.: Duke University Press, 2012), 164.
23. Harris, *Building Lives*, 135.
24. Kevin Lynch, *What Time Is This Place?* (Cambridge, Mass.: MIT Press, 1971), 178.
25. Suffolk Coastal District Council, "Report for Delegated Planning Application, DC/14/0206/FUL," March 17, 2014.
26. National Trust, *Notice of Members' Annual General Meeting, 2014.*
27. Webcast, "Annual General Meeting," *National Trust*, November 8, 2014, http://nationaltrust.vualto.com/2014/archive/morning -session/.
28. National Trust, *Notice of Members' Annual General Meeting, 2014*, "Members' Resolution on Listed Buildings," 18.
29. Ibid., "Members' Resolution on Coastal Properties, Climate Change and Community Consultation," 19.
30. Webcast, "Annual General Meeting," http://nationaltrust.vualto .com/2014/archive/morning-session/.
31. National Trust, *Notice of Members' Annual General Meeting, 2014*, "Board of Trustees Response to the Resolutions about Engaging with Local Communities," 20.
32. Phil Dyke, National Trust coast and marine advisor, personal communication , November 19, 2014.
33. These issues were discussed in early consultations between English Heritage, the National Trust, and Trinity House. Phil Dyke, maintains that if the organization had taken over management of the structure in 2013, it would have been possible to experiment with ways of creatively recycling elements of the structure while gradually dismantling it.
34. For a discussion of the way that the concept of the Anthropocene relates to framings of the past, see the special forum,

"Archaeology of the Anthropocene," *Journal of Contemporary Archaeology* 1, no. 1 (2014).

35. Joan Iverson Nassauer and Julie Raskin, "Urban Vacancy and Land Use Legacies," *Landscape and Urban Planning* 125 (2014): 245–53.

36. Steve Hinchliffe, "Reconstituting Nature Conservation: Towards a Careful Political Ecology," *Geoforum* 39, no. 1 (2008): 88–97, 95.

37. Seung-Jin Chung, "East Asian Values in Historic Conservation," *Journal of Architectural Conservation* 1 (2005): 55–70; Nobuto Ito, "'Authenticity' Inherent in Cultural Heritage in Asia and Japan," paper presented at Nara Conference on Authenticity, in *Proceedings of the UNESCO World Heritage Convention*, 1994 (Tokyo, Japan: Agency for Cultural Affairs; UNESCO World Heritage Centre, 1995).

38. Although the framing of this dynamic within heritage discourse as uniquely Eastern is in part born out of a resistance to imported technical and scientific heritage orthodoxies, see Tim Winter, "Beyond Eurocentrism? Heritage Conservation and the Politics of Difference," *International Journal of Heritage Studies* 20, no. 2 (2014): 123–37.

39. Andrew Juniper, *Wabi-Sabi: The Japanese Art of Impermanence* (Tokyo: Tuttle Publishing, 2003), 51; see also Leonard Koren, *Wabi-Sabi for Artists, Designers, Poets, and Philosophers* (Berkeley, Calif.: Stone Bridge Press, 1994).

40. J. B. Gillette, "On Her Own Terms," *Historic Preservation*, November–December 1992, cited in Lowenthal, "Value of Age and Decay," 45.

41. David Lowenthal, "Material Preservation and Its Alternatives," *Perspecta* 25 (1989): 77.

42. Ito, "Authenticity."

43 Siân Jones, "The Growth of Things and the Fossilisation of Heritage," in *A Future for Archaeology: The Past in the Present*, ed. Robert Layton, Stephen Shennan, and Peter Stone (London: UCL Press, 2006), 107–26.

44. Sven Ouzman, "The Beauty in Letting Go: Fragmentary Museums and the Archaeologies of Archive," in *Sensible Objects: Colonialism, Museums, and Material Culture*, ed. Elizabeth Edwards, Chris Gosden, and Ruth B. Philips (Oxford: Berg, 2006).

45. José Luis Romanillos, "Mortal Questions: Geographies on the Other Side of Life," *Progress in Human Geography* 39, no. 5 (2015): 566.

46. Ibid., 8.

47. Ann Laura Stoler, *Imperial Debris: On Ruins and Ruination* (Durham, N.C.: Duke University Press, 2013).

48. Paul Harrison, "Corporeal Remains: Vulnerability, Proximity, and Living On after the End of the World," *Environment and Planning A* 40 (2008): 426.

49. Jane M. Jacobs and Peter Merriman, "Practising Architectures," *Social and Cultural Geography* 12, no. 3 (2013): 212.

50. Paul Kingsnorth, "Upon the Mathematics of Falling Away," *Dark Mountain* 2 (2011): 60.

51. Duncan Kent, formerly National Trust senior ranger at Orford Ness, was instrumental in promoting artistic engagement with the loss of the lighthouse. He worked closely with Simon Read and Liz Ferretti, as well as offering his support to broader community initiatives.

52. Liz Ferretti, "The Orfordness Lighthouse Project," *EADT Suffolk Magazine*, 2015.

53. Liz Ferretti, personal communication, April 29, 2015.

54. Elisabeth Kübler-Ross and David Kesslar, *On Grief and Grieving: Finding the Meaning of Grief through the Five Stages of Loss* (New York: Scribner, 2007).

55. Richard J. Hobbs, "Grieving for the Past and Hoping for the Future: Balancing Polarizing Perspectives in Conservation and Restoration," *Restoration Ecology* 21, no. 2 (2013): 145–48.

56. Ibid., 148.

57. Kathryn Yusoff, "Aesthetics of Loss: Biodiversity, Banal Violence, and Biotic Subjects," *Transactions of the Institute of British Geographers* 37 (2011): 578–92, 579.

58. Peter Greenaway, writer and director, *A Zed and Two Noughts* (1985).

59. Tim Ingold, "No More Ancient, No More Human: The Future Past of Archaeology and Anthropology," in *Archaeology and Anthropology*, ed. Duncan Garrow and Thomas Yarrow (Oxford: Oxbow Books, 2010), 163–64.

60. Ibid., 164.

61. Tim Cresswell and Gareth Hoskins, "Place, Persistence, and Practice: Evaluating Historical Significance at Angel Island, San Francisco, and Maxwell Street, Chicago," *Annals of the Association of American Geographers* 98, no. 2 (2008): 392–413.

62. The National Trust was in conversation with Trinity House about salvaging elements of the lighthouse before its decommissioning.

63. Yusoff, "Aesthetics of Loss," 590.

64. Jens Weinberg, "Four Churches and a Lighthouse: Preservation, 'Creative Dismantling,' or Destruction," *Danish Journal of Archaeology* 3, no. 1 (2014): 68–75.
65. Tim Flohr Sørensen, "Transience and the Objects of Heritage: A Matter of Time," *Danish Journal of Archaeology* 3, no. 1 (2014): 86–90.
66. Yve-Alain Bois and Rosalind Krauss, "A User's Guide to Entropy," *October* 78 (1996): 57.
67. Robert Hobbs, *Robert Smithson: Sculpture* (Ithaca, N.Y.: Cornell University Press, 1981), 185–86.
68. Bois and Krauss, "User's Guide," 58.
69. Jorge Otero-Pailos, "Creative Agents," *Future Anterior* 3, no. 1 (2006): iii–vii.
70. Robert Smithson, "Entropy Made Visible," in *Robert Smithson: Collected Writings*, ed. Jack Flam (Berkeley: University of California Press, 1996), 307.
71. I'm grateful to artist Bridget McKenzie for sharing the photo that documents this detail. See also McKenzie's self-published 2009 volume *Tide Clock: English Coasts and Rising Seas*.
72. Benjamin Morris, "In Defence of Oblivion: The Case of Dunwich, Suffolk," *International Journal of Heritage Studies* 20, no. 2 (2014): 196–216.
73. Romanillos, "Mortal Questions," 15.
74. Morris, "In Defence of Oblivion"; Walter Benjamin, "The Storyteller," in *Illuminations* (London: Pimlico, 1999), 86.

8. Beyond Saving

1. Chris Ford, personal communication, Grant-Kohrs Ranch, Deer Lodge, Montana, September 10, 2002.
2. Cornelius Holtorf and Oscar Ortman, "Endangerment and the Conservation Ethos in Natural and Cultural Heritage: The Case of Zoos and Archaeological Sites," *International Journal of Heritage Studies* 14, no. 1 (2008): 86.
3. Cornelius Holtorf, "Preservation Paradigm in Heritage Management," in *Encyclopedia of Global Archaeology*, ed. Claire Smith (New York: Springer, 2014), 6128–31.
4. Françoise Choay, *The Invention of the Historic Monument*, trans. L. M. O'Connell (Cambridge: Cambridge University Press, 2001).
5. Cornelius Holtorf and Anders Höberg, "Contemporary Heri-

tage and the Future," in *Palgrave Handbook of Contemporary Heritage Research*, ed. Emma Waterton and Steve Watson (London: Palgrave Macmillan, 2015), 514.

6. Holtorf and Ortman, "Endangerment," 87.

7. Greg Kennedy, *An Ontology of Trash: The Disposable and Its Problematic Nature* (Albany: State University of New York Press, 2007), 136.

8. Ibid., 131.

9. Ibid., 147.

10. Ibid., 135.

11. Castle Drogo, *National Trust*, http://www.nationaltrust.org.uk/castle-drogo.

12. National Trust, "Castle Drogo Outside In: A Creative Opportunity," call for expressions of interest (unpublished, 2013).

13. *The Outside In Room* was created by Penny Saunders and Tim Britton of Forkbeard Fantasy, in association with creative partners Mdesign. *The Four-Poster Bed* and the *Little Cupboard of Decay* were built by Tim Rae-Duke.

14. Aron Vinegar and Jorge Otero-Pailos, "What a Monument Can Do," *Future Anterior* 8, no. 2 (2011): vii.

15. Ibid.

16. Elizabeth Grosz, *Architecture from the Outside: Essays on Virtual and Real Space* (Cambridge, Mass.: MIT Press, 2001), 91.

17. Michael E. Zimmerman, "Heidegger, Buddhism, and Deep Ecology," in *Cambridge Companion to Heidegger* (Cambridge: Cambridge University Press, 2006), 295–96.

18. Ibid., 263.

19. Rodney Harrison, "Beyond 'Natural' and 'Cultural' Heritage: Towards an Ontological Politics of Heritage in the Age of the Anthropocene," *Heritage and Society* 8, no. 1 (2015): 32.

20. Deborah Bird Rose, *Wild Dog Dreaming: Love and Extinction* (Charlottesville: University of Virginia Press, 2011), 143, cited in Harrison, "Beyond 'Natural' and 'Cultural,'" 32.

21. Ioannis Poulios, "Moving Beyond a Values-Based Approach to Heritage Conservation," *Conservation and Management of Archaeological Sites* 12, no. 2 (2010): 175.

22. Ibid.

23. Anna Lydford, "Clever Use of Recycling Rebuilds Tremayne Quay," *Constant Times* 5, no. 2 (2016): 4.

24. Kevin Lynch, *What Time Is This Place?* (Cambridge, Mass.: MIT Press, 1972), 53.

25. John Wylie, "Landscape, Absence, and the Geographies of Love," *Transactions of the Institute of British Geographers* 34, no. 3 (2009): 275–89.

26. Alois Riegl, "The Modern Cult of Monuments," in *Historical and Philosophical Issues in the Conservation of Cultural Heritage*, ed. N. S. Price, M. K. Talley, and A. M. Vaccaro (1903; repr., Los Angeles, Calif.: Getty Conservation Institute, 1996), 69–83.

27. Mats Burström, "Fragments as Something More: Archaeological Experience and Reflection," in *Reclaiming Archaeology: Beyond the Tropes of Modernity*, ed. Alfredo González-Ruibal (London: Routledge, 2013), 313.

28. Ibid., 319.

29. Jane M. Jacobs and Stephen Cairns, *Buildings Must Die: A Perverse View of Architecture* (Cambridge, Mass.: MIT Press, 2014), 187.

Permissions

Index

Caitlin DeSilvey is associate professor of cultural geography at the University of Exeter. She is coauthor of *Visible Mending* (with Steven Bond and James R. Ryan) and coeditor of *Anticipatory History* (with Simon Naylor and Colin Sackett).

Printed and bound by CPI Group (UK) Ltd, Croydon, CR0 4YY

27/10/2024

14580403-0001